ASE Test Preparation Series

Automobile Test

Advanced Engine Performance (Test L1)

4th Edition

DELMAR
CENGAGE Learning

Australia • Brazil • Japan • Korea • Mexico • Singapore • Spain • United Kingdom • United States

DELMAR
CENGAGE Learning

ASE Test Preparation Series: Automobile Test for Advanced Engine Performance (Test L1), Fourth Edition

Vice President, Technology Professional Business Unit: Gregory L. Clayton

Product Development Manager: Kristen Davis

Product Manager: Kim Blakey

Editorial Assistant: Vanessa Carlson

Director of Marketing: Beth A. Lutz

Marketing Specialist: Brian McGrath

Marketing Coordinator: Marissa Maiella

Production Manager: Andrew Crouth

Production Editor: Kara A. DiCaterino

Senior Project Editor: Christopher Chien

XML Architect: Jean Kaplansky

Cover Design: Michael Egan

Cover Images: Portion courtesy of DaimlerChrysler Corporation

For product information and technology assistance, contact us at
Cengage Learning Customer & Sales Support, 1-800-354-9706

For permission to use material from this text or product, submit all requests online at **www.cengage.com/permissions**
Further permissions questions can be emailed to
permissionrequest@cengage.com

ISBN-13: 978-1-4180-3888-5

ISBN-10: 1-4180-3888-1

Delmar
Executive Woods
5 Maxwell Drive
Clifton Park, NY 12065
USA

Cengage Learning is a leading provider of customized learning solutions with office locations around the globe, including Singapore, the United Kingdom, Australia, Mexico, Brazil, and Japan. Locate your local office at **international.cengage.com/region**

Cengage Learning products are represented in Canada by Nelson Education, Ltd.

For your lifelong learning solutions, visit **delmar.cengage.com**

Visit our corporate website at **www.cengage.com**

Notice to the Reader

Printed in the United States of America
5 6 7 11 10 09

Contents

Section   1   The History and Purpose of ASE

Section   2   Take and Pass Every ASE Test

Section   3   Types of Questions on an ASE Exam

Section   4   An Overview of the Task List

Section 5 Sample Test for Practice

Section 6 Additional Test Questions for Practice

Section 7 Appendices

Preface

Delmar Cengage Learning is very pleased that you have chosen our ASE Test Preparation Series to prepare yourself for the automotive ASE Examination. These guides are available for all of the automotive areas including A1–A8, the L1 Advanced Diagnostic Certification, the P2 Parts Specialist, the C1 Service Consultant and the X1 Undercar Specialist. These guides are designed to introduce you to the Task List for the test you are preparing to take, give you an understanding of what you are expected to be able to do in each task, and take you through sample test questions formatted in the same way the ASE tests are structured.

If you have a basic working knowledge of the discipline you are testing for, you will find Delmar's ASE Test Preparation Series to be an excellent way to understand the "must know" items to pass the test. These books are not textbooks. Their objective is to prepare the technician who has the requisite experience and schooling to challenge ASE testing. It cannot replace the hands-on experience or the theoretical knowledge required by ASE to master vehicle repair technology. If you are unable to understand more than a few of the questions and their explanations in this book, it could be that you require either more shop-floor experience or further study. Some resources that can assist you with further study are listed on the rear cover of this book.

Each book begins with an item-by-item overview of the ASE Task List with explanations of the minimum knowledge you must possess to answer questions related to the task. Following that there are 2 sets of sample questions followed by an answer key to each test and an explanation of the answers to each question. A few of the questions are not strictly ASE format but were included because they help teach a critical concept that will appear on the test. We suggest that you read the complete Task List Overview before taking the first sample test. After taking the first test, score yourself and read the explanation to any questions that you were not sure about, including the questions you answered correctly. Each test question has a reference back to the related task or tasks that it covers. This will help you to go back and read over any area of the task list that you are having trouble with. Once you are satisfied that you have all of your questions answered from the first sample test, take the additional tests and check them. If you pass these tests, you will be prepared to do well on the ASE test.

Our Commitment to Excellence

The 4th edition of Delmar's ASE Test Preparation Series has been through a major revision with extensive updates to the ASE's task lists, test questions, and answers and explanations. Delmar Cengage Learning has sought out the best technicians in the country to help with the updating and revision of each of the books in the series.

About the Series Advisor

To promote consistency throughout the series, a series advisor took on the task of reading, editing, and helping each of our experts give each book the highest level of accuracy possible. Dan Perrin has served in the role of Series Advisor for the 4th edition of the ASE Test Preparation Series. Dan began ASE testing with the first series of tests in 1972 and has been continually certified ever since. He holds ASE master status in automotive, truck, collision, and machinist. He is also L1, L2, and alternated fuels certified, along with some others that have expired. He has been an automotive educator since 1979, having taught at the secondary, post-secondary, and industry levels. His service includes participation on boards that include the North American Council of Automotive Teachers (NACAT), the Automotive Industry Planning Council (AIPC), and the National Automotive Technicians Education Foundation (NATEF). Dan currently serves as the Executive Manager of NACAT and Director of the NACAT Education Foundation.

Thanks for choosing Delmar's ASE Test Preparation Series. All of the writers, editors, Delmar Staff, and myself have worked very hard to make this series second to none. I know you are going to find this book accurate and easy to work with. It is our objective to constantly improve our product at Delmar by responding to feedback.

If you have any questions concerning the books in this series, you can email me at: autoexpert@trainingbay.com.

Dan Perrin
Series Advisor

1 The History and Purpose of ASE

ASE began as the National Institute for Automotive Service Excellence (NIASE). It was founded as a non-profit independent entity in 1972 by a group of industry leaders with the single goal of providing a means for consumers to distinguish between incompetent and competent technicians. It accomplishes this goal by testing and certification of repair and service professionals. From this beginning it has evolved to be known simply as ASE (Automotive Service Excellence) and today offers more than 40 certification exams in automotive, medium/heavy duty truck, collision, engine machinist, school bus, parts specialist, automobile service consultant, and other industry-related areas. At this time there are more than 400,000 professionals with current ASE certifications. These professionals are employed by new car and truck dealerships, independent garages, fleets, service stations, franchised service facilities, and more. ASE continues its mission by also providing information that helps consumers identify repair facilities that employ certified professionals through its Blue Seal of Excellence Recognition Program. Shops that have a minimum of 75% of their repair technicians ASE certified and meet other criteria can apply for and receive the Blue Seal of Excellence Recognition from ASE.

ASE recognized that educational programs serving the service and repair industry also needed a way to be recognized as having the faculty, facilities, and equipment to provide a quality education to students wanting to become service professionals. Through the combined efforts of ASE, industry, and education leaders, the non-profit National Automotive Technicians Education Foundation (NATEF) was created to evaluate and recognize training programs. Today more than 2000 programs are ASE certified under the standards set by the service industry. ASE/NATEF also has a certification of industry (factory) training program known as CASE. CASE stands for Continuing Automotive Service Education and recognizes training provided by replacement parts manufacturers as well as vehicle manufacturers.

ASE certification testing is administered by the American College Testing (ACT). Strict standards of security and supervision at the test centers insure that the technician who holds the certification earned it. Additionally ASE certification also requires that the person passing the test to be able to demonstrate that they have two years of work experience in the field before they can be certified. Test questions are developed by industry experts that are actually working in the field being tested. There is more detail on how the test is developed and administered in the next section. Paper and pencil tests are administered twice a year at over seven hundred locations in the United States. Computer based testing is now also available with the benefit of instant test results at certain established test centers. The certification is valid for five years and can be recertified by retesting. So that consumers can recognize certified technicians, ASE issues a jacket patch, certificate, and wallet card to certified technicians and makes signs available to facilities that employ ASE certified technicians.

You can contact ASE at any of the following:

National Institute for Automotive Service Excellence
101 Blue Seal Drive S.E.
Suite 101
Leesburg, VA 20175
Telephone 703-669-6600
FAX 703-669-6123
www.ase.com

2 Take and Pass Every ASE Test

Participating in an Automotive Service Excellence (ASE) voluntary certification program gives you a chance to show your customers that you have the "know-how" needed to work on today's modern vehicles. The ASE certification tests allow you to compare your skills and knowledge to the automotive service industry's standards for each specialty area.

If you are the "average" automotive technician taking this test, you are in your mid-thirties and have not attended school for about fifteen years. That means you probably have not taken a test in many years. Some of you, on the other hand, have attended college or taken postsecondary education courses and may be more familiar with taking tests and with test-taking strategies. There is, however, a difference in the ASE test you are preparing to take and the educational tests you may be accustomed to.

How are the tests administered?

ASE test are administered at over 750 test sites in local communities. Paper and pencil tests are the type most widely available to technicians. Each tester is given a booklet containing questions with charts and diagrams where required. You can mark in this test booklet but no information entered in the booklet is scored. Answers are recorded on a separate answer sheet. You will enter your answers, using a number 2 pencil only. ASE recommends you bring four sharpened number 2 pencils that have erasers. Answer choices are recorded by coloring in the blocks on the answer sheet. The answer sheets are scanned electronically and the answers tabulated. For test security, test booklets include randomly generated questions. Your answer key must be matched to the proper booklet so it is important to correctly enter the booklet serial number on the answer sheet. All instructions are printed on the test materials and should be followed carefully.

ASE has introduced Computer Based Testing (CBT) at some locations. While the test content is the same for both testing methods the CBT tests have some unique requirements and advantages. It is strongly recommended that technicians considering the CBT tests go the ASE web page at www.ASE.com and review the conditions and requirements for this type of test. There is a demonstration of a CBT that allows you to experience this type of test before you register. Some technicians find this style of testing provides an advantage, while others find operating the computer a distraction. One significant benefit of CBT is the availability of instant results. You can receive your test results before you leave the test center. CBT testing also offers increased flexibility in scheduling. The cost for taking CBTs is slightly higher than paper and pencil tests and the number of testing sites is limited. The first time test taker may be more comfortable with the paper and pencil tests but technicians now have a choice.

Who Writes the Questions?

The questions are written by service industry experts in the area being tested. Each area will have its own technical experts. Questions are entirely job related. They are designed to test the skills you need to be a successful technician. Theoretical knowledge is important and necessary to answer the questions, but the ability to apply that knowledge is the basis of ASE test questions.

Each question has its roots in an ASE "item-writing" workshop where service representatives from automobile manufacturers (domestic and import), aftermarket parts and equipment manufacturers, working technicians, and vocational educators meet in a workshop setting to share ideas and translate them into test questions. Each test question written by these experts must survive review by all members of the group.

The questions are written to deal with practical application of soft skills and system knowledge experienced by technicians in their day-to-day work.

All questions are pre-tested and quality-checked on a national sample of technicians. Those questions that meet ASE standards of quality and accuracy are included in the scored sections of the tests; the "rejects" are sent back to the drawing board or discarded altogether.

Each certification test is made up of between forty and eighty multiple-choice questions.

Note: Each test could contain additional questions that are included for statistical research purposes only. Your answers to these questions will not affect your score, but since you do not know which ones they are, you should answer all questions on the test. The five-year Recertification Test will cover the same content areas as those listed above. However, the number of questions in each content area of the Recertification Test will be reduced by about one-half.

Objective Tests

A test is called an objective test if the same standards and conditions apply to everyone taking the test and there is only one correct answer to each question.

Objective tests primarily measure your ability to recall information. A well-designed objective test can also test your ability to understand, analyze, interpret, and apply your knowledge. Objective tests include true-false, multiple choice, fill in the blank, and matching questions. ASE's tests consist exclusively of four-part multiple-choice objective questions.

The following are some strategies that may be applied to your tests.

Before beginning to take an objective test, quickly look over the test to determine the number of questions, but do not try to read through all of the questions. In an ASE test, there are usually between forty and eighty questions, depending on the subject. Read through each question before marking your answer. Answer the questions in the order they appear on the test. Leave the questions blank that you are not sure of and move on to the next question. You can return to those unanswered questions after you have finished the others. They may be easier to answer at a later time after your mind has had additional time to consider them on a subconscious level. In addition, you might find information in other questions that will help you recall the answers to some of them.

Do not be obsessed by the apparent pattern of responses. For example, do not be influenced by a pattern like **D, C, B, A, D, C, B, A** on an ASE test.

There is also a lot of folk wisdom about taking objective tests. For example, there are those who would advise you to avoid response options that use certain words such as *all, none, always, never, must,* and *only,* to name a few. This, they claim, is because nothing in life is exclusive. They would advise you to choose response options that use words that allow for some exception, such as *sometimes, frequently, rarely, often, usually, seldom,* and *normally.* They would also advise you to avoid the first and last option (A and D) because test writers, they feel, are more comfortable if they put the correct answer in the middle (B and C) of the choices. Another recommendation often offered is to select the option that is either shorter or longer than the other three choices because it is more likely to be correct. Some would advise you to never change an answer since your first intuition is usually correct.

Although there may be a grain of truth in this folk wisdom, ASE test writers try to avoid them and so should you. There are just as many **A** answers as there are **B** answers, just as many **D** answers as **C** answers. As a matter of fact, ASE tries to balance the answers at about 25 percent per choice **A, B, C,** and **D.** There is no intention to use "tricky" words, such as outlined above. Put no credence in the opposing words "sometimes" and "never," for example.

Multiple-choice tests are sometimes challenging because there are often several choices that may seem possible, and it may be difficult to decide on the correct choice. The best strategy, in this case, is to first determine the correct answer before looking at the options. If you see the answer you decided on, you should still examine the options to make sure that none seem more correct than yours. If you do not know or are not sure of the answer, read each option very carefully and try to eliminate those options that you know to be wrong. That way, you can often arrive at the correct choice through a process of elimination.

If you have gone through all of the test and you still do not know the answer to some of the questions, <u>then guess.</u> Yes, guess. You then have at least a 25 percent chance of being correct. If you leave the question blank, you have no chance. Your score is based on the number of questions answered correctly.

Preparing for the Exam

The main reason we have included so many sample and practice questions in this guide is, simply, to help you learn what you know and what you don't know. We recommend that you work your way through each question in this book. Before doing this, carefully look through Section 3; it contains a description and explanation of the question types you'll find on an ASE exam.

Once you understand what the questions will look like, move to the sample test. Answer one of the sample questions (Section 5) then read the explanation (Section 7) to the answer for that question. If you don't feel you understand the reasoning for the correct answer, go back and read the overview (Section 4) for the task that is related to that question. If you still don't feel you have a solid understanding of the material, identify a good source of information on the topic, such as a textbook, and do some more studying.

After you have completed all of the sample test items and reviewed your answers, move to the additional questions (Section 6). This time answer the questions as if you were taking an actual test. Do not use any reference or allow any interruptions in order to get a feel for how you will do on an actual test. Once you have answered all of the questions, grade your results using the answer key in Section 7. For every question that you gave a wrong answer to, study the explanations to the answers and/or the overview of the related task areas. Try to determine the root cause for your missing the question. The easiest thing to correct is learning the correct technical content. The hardest thing to correct is behaviors that lead you to a wrong conclusion. If you knew the information but still got it wrong there is a behavior problem that will need to be corrected. An example would be reading too quickly and skipping over words that affect your reasoning. If you can identify what you did that caused you to answer the question incorrectly you can eliminate that cause and improve your score. Here are some basic guidelines to follow while preparing for the exam:

- Focus your studies on those areas you are weak in.

- Be honest with yourself while determining if you understand something.

- Study often but in short periods of time.

- Remove yourself from all distractions while studying.

- Keep in mind the goal of studying is not just to pass the exam, the real goal is to learn!

- Prepare physically by getting a good night's rest before the test and eat meals that provide energy but do not cause discomfort.

- Arrive early to the test site to avoid long waits as test candidates check in and to allow all of the time available for your tests.

During the Test

On paper and pencil tests you will be placing your answers on a sheet where you will be required to color in your answer choice. Stray marks or incomplete erasures may be picked up as an answer by the electronic reader, so be sure only your answers end up on the sheet. One of the biggest problems an adult faces in test taking, it seems, is placing the answer in the correct spot on the answer sheet. Make certain that you mark your answer for, say, question 21, in the space on the answer sheet designated for the answer for question 21. A correct response in the wrong line will probably result in two questions being marked wrong, one with two answers (which could include a correct answer but will be scored wrong) and the other with no answer. Remember, the answer sheet on the written test is machine scored and can only "read" what you have colored in.

If you finish answering all of the questions on a test and have remaining time, go back and review the answers to those questions that you were not sure of. You can often catch careless errors by using the remaining time to review your answers. Carefully check your answer sheet for blank answer blocks or missing information.

At practically every test, some technicians will invariably finish ahead of time and turn their papers in long before the final call. Some technicians may be doing recertification tests and others may be taking fewer tests than you. Do not let them distract or intimidate you.

It is not wise to use less than the total amount of time that you are allotted for a test. If there are any doubts, take the time for review. Any product can usually be made better with some additional effort. A test is no exception. It is not necessary to turn in your test paper until you are told to do so.

Testing Time Length

An ASE written test session is four hours. You may attempt from one to a maximum of four tests in one session. It is recommended, however, that no more than a total of 225 questions be attempted at any test session. This will allow for just over one minute for each question.

Visitors are not permitted at any time. If you wish to leave the test room, for any reason, you must first ask permission. If you finish your test early and wish to leave, you are permitted to do so only during specified dismissal periods.

You should monitor your progress and set an arbitrary limit to how much time you will need for each question. This should be based on the number of questions you are attempting. It is suggested that you wear a watch because some facilities may not have a clock visible to all areas of the room.

Computer-Based Tests are allotted a testing time according to the number of questions ranging from one half hour to one and one half hours. Advanced level tests are allowed two hours. This time is by appointment and you should be sure to be on time to insure that you have all of the time allocated. If you arrive late for a CBT test appointment you will only have the amount of time remaining on your appointment.

Your Test Results!

You can gain a better perspective about tests if you know and understand how they are scored. ASE's tests are scored by American College Testing (ACT), a nonpartial, unbiased organization having no vested interest in ASE or in the automotive industry.

Each question carries the same weight as any other question. For example, if there are fifty questions, each is worth 2 percent of the total score. The passing grade is 70 percent. That means you must correctly answer thirty-five of the fifty questions to pass the test.

The test results can tell you:

- where your knowledge equals or exceeds that needed for competent performance, or

- where you might need more preparation.

Your ASE test score report is divided into content areas and will show the number of questions in each content area and how many of your answers were correct. These numbers provide information about your performance in each area of the test. However, because there may be a different number of questions in each content area of the test, a high percentage of correct answers in an area with few questions may not offset a low percentage in an area with many questions.

It should be noted that one does not "fail" an ASE test. The technician who does not pass is simply told "More Preparation Needed." Though large differences in percentages may indicate problem areas, it is important to consider how many questions were asked in each area. Since each test evaluates all phases of the work involved in a service specialty, you should be prepared in each area. A low score in one area could keep you from passing an entire test.

There is no such thing as average. You cannot determine your overall test score by adding the percentages given for each task area and dividing by the number of areas. It doesn't work that way because there generally are not the same number of questions in each task area. A task area with twenty questions, for example, counts more toward your total score than a task area with ten questions.

Your test report should give you a good picture of your results and a better understanding of your strengths and weaknesses for each task area.

If you fail to pass the test, you may take it again at any time it is scheduled to be administered. You are the only one who will receive your test score. Test scores will not be given over the telephone by ASE nor will they be released to anyone without your written permission.

3 Types of Questions on an ASE Exam

ASE certification tests are often thought of as being tricky. They may seem to be tricky if you do not completely understand what is being asked. The following examples will help you recognize certain types of ASE questions and avoid common errors.

Paper-and-pencil tests and computer-based test questions are identical in content and difficulty. Most initial certification tests are made up of forty to eighty multiple-choice questions. Multiple-choice questions are an efficient way to test knowledge. To answer them correctly, you must think about each choice as a possibility, and then choose the one that best answers the question. To do this, read each word of the question carefully. Do not assume you know what the question is about until you have finished reading it.

About 10 percent of the questions on an actual ASE exam will use an illustration. These drawings contain the information needed to correctly answer the question. The illustration must be studied carefully before attempting to answer the question. Often, technicians look at the possible answers then try to match up the answers with the drawing. Always do the opposite; match the drawing to the answers. When the illustration is showing an electrical schematic or another system in detail, look over the system and try to figure out how the system works before you look at the question and the possible answers.

Multiple-Choice Questions

The most common type of question used on ASE Tests is the multiple-choice question. This type of question contains three "distracters" (wrong answers) and one "key" (correct answer). When the questions are written effort is made to make the distracters plausible to draw an inexperienced technician to one of them. This type of question gives a clear indication of the technician's knowledge. Using multiple criteria including cross-sections by age, race, and other background information, ASE is able to guarantee that a question does not bias for or against any particular group. A question that shows bias toward any particular group is discarded. If you encounter a question that you are unsure of, reverse engineer it by eliminating the items that it cannot be. For example:

A rocker panel is a structural member of which vehicle construction type?

A. Front-wheel drive
B. Pickup truck
C. Unibody
D. Full-frame

Analysis:

This question asks for a specific answer. By carefully reading the question, you will find that it asks for a construction type that uses the rocker panel as a structural part of the vehicle.

Answer A is wrong. Front-wheel drive is not a vehicle construction type.
Answer B is wrong. A pickup truck is not a type of vehicle construction.
Answer C is correct. Unibody design creates structural integrity by

welding parts together, such as the rocker panels, but does not require exterior cosmetic panels installed for full strength.

Therefore, the correct answer is C. If the question was read quickly and the words "construction type" were passed over, answer A may have been selected.

Answer D is wrong. Full-frame describes a body-over-frame construction type that relies on the frame assembly for structural integrity.

Therefore, the correct answer is C. If the question was read quickly and the words "construction type" were passed over, answer A may have been selected.

EXCEPT Questions

Another type of question used on ASE tests has answers that are all correct except one. The correct answer for this type of question is the answer that is wrong. The word "**EXCEPT**" will always be in capital letters. You must identify which of the choices is the wrong answer. If you read quickly through the question, you may overlook what the question is asking and answer the question with the first correct statement. This will make your answer wrong. An example of this type of question and the analysis is as follows:

All of the following are tools for the analysis of structural damage **EXCEPT:**

A. height gauge
B. tape measure.
C. dial indicator.
D. tram gauge.

Analysis:

The question really requires you to identify the tool that is not used for analyzing structural damage. All tools given in the choices are used for analyzing structural damage except one. This question presents two basic problems for the test-taker who reads through the question too quickly. It may be possible to read over the word "**EXCEPT**" in the question or not think about which type of damage analysis would use answer C. In either case, the correct answer may not be selected. To correctly answer this question, you should know what tools are used for the analysis of structural damage. If you cannot immediately recognize the incorrect tool, you should be able to identify it by analyzing the other choices.

Answer A is wrong. A height gauge may be used to analyze structural damage.
Answer B is wrong. A tape measure may be used to analyze structural damage.
Answer C is correct. A dial indicator may be used as a damage analysis tool for moving parts, such as wheels, wheel hubs, and axle shafts, but would not be used to measure structural damage.
Answer D is wrong. A tram gauge is used to measure structural damage.

Technician A, Technician B Questions

The type of question that is most popularly associated with an ASE test is the "Technician A says . . . Technician B says . . . Who is right?" type. In this type of question, you must identify the correct statement or statements. To answer this type of question correctly, you must carefully read each technician's statement and judge it on its own merit to determine if the statement is true.

Sometimes this type of question begins with a statement about some analysis or repair procedure. This is often referred to as the stem of the question and provides the setup or background information required to understand the conditions the question is based on. This is followed by two statements about the cause of the concern, proper inspection, identification, or repair choices. You are asked whether the first statement, the second statement, both statements, or neither statement is correct.

Analyzing this type of question is a little easier than the other types because there are only two ideas to consider although there are still four choices for an answer.

Technician A, Technician B questions are really double true or false questions. The best way to analyze this kind of question is to consider each technician's statement separately. Ask yourself, is A true or false? Is B true or false? Then select your answer from the four choices. An important point to remember is that an ASE Technician A, Technician B question will never have Technician A and B directly disagreeing with each other. That is why you must evaluate each statement independently.

An example of this type of question and the analysis of it follows.

A vehicle comes into the shop with a gas gauge that will not register above one half full. When the sending unit circuit is disconnected the gauge reads empty and when it is connected to ground the gauge goes to full. Technician A says that the sending unit is shorted to ground. Technician B says the gauge circuit is working and the sending unit is likely the problem. Who is right?

A. A only
B. B only
C. Both A and B
D. Neither A nor B

Analysis:

Reading of the stem of the question sets the conditions of the customer concern and establishes what information is gained from testing. General knowledge of gauge circuits and test procedures are needed to correctly evaluate the technician's conclusions. Note: Avoid being distracted by experience with unusual or problem vehicles that you may have worked on, Other technicians taking the same test do not have that knowledge, so it should not be used as the basis of your answers.

Technician A is wrong because a shorted to ground sending unit would produce a gauge reading equivalent to the test conditions of a grounding the circuit and produce a full reading. **Technician B is correct** because the gauge spans when going from an open circuit to a completely
grounded circuit. This would tend to indicate that the problem had to be in the sending unit. Answer C is not correct. Both technicians are identifying the problem as a sending unit but technician A qualified the problem as a specific type of failure (grounded) that would not have caused the symptoms of the vehicle.
Answer D is not correct because technician B's diagnosis is a possible cause of the conditions identified.

Most-Likely Questions

Most-Likely questions are somewhat difficult because only one choice is correct while the other three choices are nearly correct. An example of a Most-Likely-cause question is as follows:

The Most-Likely cause of reduced turbocharger boost pressure may be a:

A. wastegate valve stuck closed.
B. wastegate valve stuck open.
C. leaking wastegate diaphragm.
D. disconnected wastegate linkage.

Analysis:

Answer A is wrong. A wastegate valve stuck closed increases turbocharger boost pressure.

Answer B is correct. A wastegate valve stuck open decreases turbocharger boost pressure.
Answer C is wrong. A leaking wastegate valve diaphragm increases turbocharger boost pressure.
Answer D is wrong. A disconnected wastegate valve linkage will increase turbocharger boost pressure.

LEAST-Likely Questions

Notice that in Most-Likely questions there is no capitalization. This is not so with LEAST-Likely type questions. For this type of question, look for the choice that would be the LEAST-Likely cause of the described situation. Read the entire question carefully before choosing your answer. An example is as follows:

What is the LEAST-Likely cause of a bent pushrod?

A. Excessive engine speed
B. A sticking valve
C. Excessive valve guide clearance
D. A worn rocker arm stud

Analysis:

Answer A is wrong. Excessive engine speed may cause a bent pushrod.
Answer B is wrong. A sticking valve may cause a bent pushrod.
Answer C is correct. Excessive valve clearance will not generally cause a bent pushrod.
Answer D is wrong. A worn rocker arm stud may cause a bent pushrod.

You should avoid relating questions to those unusual situations that you may have encountered and answer based on the technical and mechanical possibilities.

Summary

There are no four-part multiple-choice ASE questions having "none of the above" or "all of the above" choices. ASE does not use other types of questions, such as fill-in-the-blank, completion, true-false, word-matching, or essay. ASE does not require you to draw diagrams or sketches. If a formula or chart is required to answer a question, it is provided for you. There are no ASE questions that require you to use a pocket calculator.

An Overview of the Task List

Advanced Engine Performance Specialist (Test L1)

The following section includes the task areas and task lists for this test and a written overview of the topics covered in the test.

The Task List describes the actual work you should be able to do as a technician that you will be tested on by the ASE. This is your key to the test and you should review this section carefully. We have based our sample test and additional questions upon these tasks, and the overview section will also support your understanding of the Task List. ASE advises that the questions on the test may not equal the number of tasks listed; the Task Lists tell you what ASE expects you to know how to do and be ready to be tested upon.

At the end of each question in the Sample Test and Additional Test Questions sections, a letter and number will be used as a reference back to this section for additional study. Note the following example: **E.1.**

Task List

E. Emission-Control Systems Diagnosis (10 Questions)

Task E.1   **Inspect and test for missing, modified, or tampered components.**

Example:
  1. Technician A says you can replace original emission system parts with aftermarket parts that improve performance. Technician B says you can bypass emission components to increase power output. Who is right?
     A. Technician A only
     B. Technician B only
     C. Both A and B
     D. Neither A nor B   (E.1)

Analysis:

Question #1

Answer A is wrong. Emission system components can only be replaced with parts that meet original-equipment specifications.
Answer B is wrong. Emission components cannot be bypassed under federal law.
Answer C is wrong. Neither Technician is correct.
Answer D is correct. Neither Technician is correct.

Task List and Overview

A. General Powertrain Diagnosis (5 Questions)

Task A.1   **Inspect and test for missing, modified, inoperative, or tampered powertrain mechanical components.**

Changing or bypassing mechanical components on a late-model engine not only violates Federal law; it usually degrades vehicle performance, fuel economy, and emission control. Using a component locator manual and your knowledge of a system, inspect the system for missing, damaged, or aftermarket replacement parts. Ensure that all vacuum hoses are connected correctly and free from damage. Also be sure that wiring harnesses are connected properly and free from corroded connections and other damage such as melted or chafed insulation.

Mechanical components that are often tampered with include heated air cleaners, air intake hoses, fuel-pressure regulators, fuel injectors, carburetors, vacuum switches and solenoids, ignition distributors, exhaust gas recirculation (EGR) systems, air-injection systems, intake and exhaust manifolds (replaced with unapproved aftermarket units), and other exhaust components like the catalytic converter and muffler. Also, don't overlook the restrictor in the gas tank filler neck. Altering or rendering inoperative any of the above mentioned components could change the air/fuel mixture delivery and or requirements of the engine, as well as increase emissions or reduce fuel economy.

Aftermarket intake and exhaust manifolds can adversely affect emissions and fuel economy. Aftermarket intake manifolds must contain any provisions for exhaust gas recirculation that were present on the original equipment manifold. Failure to provide EGR on an engine where it was original equipment can cause increased NO_x emissions and detonation or pinging. Similarly, aftermarket exhaust manifolds, or headers, must contain original equipment provisions for exhaust heat risers, air-injection nozzles, and oxygen sensors. As a general rule, replacement manifolds should be selected only from the carmaker's available options or from aftermarket suppliers who have obtained approval from government agencies.

Exhaust system components that have been modified in an unapproved way can also adversely affect performance and emission control. Refer to the overview for Task A.10 for more information.

Task A.2   **Locate relevant service information.**

The amount of diagnostic and service information provided for new vehicles by the manufacturers grows enormously each year. Finding the right information and recognizing what is relevant for a specific job, and what is not, is an important skill for any technician.

Diagnostic information is published in service manuals by the carmakers, and much of this information is reprinted in aftermarket manuals by independent publishers. This information includes lists of diagnostic trouble codes (DTC), test specifications, and diagnostic flow charts. Component location information, vacuum and wiring diagrams, and part numbers also are essential service data.

Almost all diagnostic information is available in printed manuals, from the carmakers and from aftermarket publishers. Much of the information also is available on microfiche and in computer-based information systems. Some service data also can be accessed on the internet and other online computer systems.

Technical service bulletins (TSB) are an important part of every carmaker's information systems. TSBs are intended primarily for dealership service departments, but many are available through aftermarket information sources. TSBs are often used to inform dealer service technicians about manufacturer-approved changes to control system calibrations, fixes for recurrent problems, running changes made to vehicles during a model year, and service manual updates. TSBs are a primary source of up-to-date information. If available, they should be searched for relevant data for most diagnostic jobs.

Task A.3   Research system operation using technical information to determine diagnostic procedure.

The need to fully understand system and component operation is essential when diagnosing or repairing today's complex vehicles. Utilizing vehicle service manuals, computerized repair information systems or technical service bulletins, the technician should research and become familiar with the systems on the vehicle being serviced. With a thorough understanding of the systems present on a particular vehicle, a diagnostic approach can be undertaken and the correct test procedures applied.

Newer systems, such as variable valve timing and intake runner controls, can lead to incorrect diagnosis of engine performance problems, such as hard starting or lack of power, if these systems are not identified and understood. Deciding on the correct diagnostic path to follow is one of the most challenging problems facing a technician servicing computerized vehicles today. A technician must often use his/her knowledge and intuition when researching technical information to determine and understand engine-management system strategy. While sometimes providing the correct solution to a problem, manufacturer's trouble trees or flow charts often fail to provide a correct diagnosis to a problem and can lead to unnecessary parts' replacement.

Flow charts do provide a valuable source of specifications for testing systems or components. Using the specifications found in flow charts, a technician can determine what minimum and maximum allowable values are when testing a component or system. This is known as specification based diagnosis and can allow the technicians experience and ingenuity to work for him. For example, if you know the goal of the powertrain control module (PCM) is to maintain a 14.7:1 air/fuel mixture to provide low emissions, but the oxygen sensor output is fixed rich at 0.9 volts, you can make the decision that things are not working as they should. A technician may follow a flow chart and not really find a problem. The flow chart concludes that the PCM must be bad and should be replaced with a known good part.

An experienced technician knows a bad PCM is rare, so instead relying only on the flow chart, he reads through the entire chart to determine what a good oxygen sensor output specification should be. If the flow chart says normal oxygen sensor output should be 0.5 volts and cycling, then the technician knows this is what needs to be achieved to restore proper system operation. The next step is to determine through five-gas analyzer testing if the mixture is indeed rich. If the gas analyzer confirms the rich condition, then the oxygen sensor output is right and the technician must then determine which inputs and outputs can cause an over-fueling condition. Once the specifications for each of the inputs and outputs are located, the technician can set out testing each one in a logical order, using his/her intuition and experience to locate the actual cause of the problem.

Applying a logical and consistent diagnostic procedure is the most important skill a technician can possess. This procedure must be followed consistently to prevent overlooking the basics, which is where many problems are often found. Failing to follow a set procedure or process that verifies the basics will lead to diagnostic failures.

Task A.4   Use appropriate diagnostic procedures based on available vehicle data and service information; determine if available information is adequate to proceed with effective diagnosis.

After service information has been gathered and reviewed, you must decide if you have enough information to proceed with accurate diagnosis and repair. Current TSB information may be essential for a car that is one or two years old, but it may not be relevant for an older model. Information on conditions that cause a certain DTC to set may be essential when troubleshooting a specific DTC, but may not be relevant if no codes are present. A wiring diagram may be essential for complicated electrical troubleshooting and repair; for some repairs, following the wire color codes on the vehicle may be sufficient.

Probably the most important piece of service information is an accurate verification of the complaint or the problem. As a diagnostic technician, you should get all possible information from the car owner or driver about drivability symptoms and about any recent work done on the car. If a symptom is present during your diagnosis and you can verify it personally, the job is often easier. If the symptom is not present, getting a complete description from the driver is even more important.

Problems that are intermittent illustrate the need for improved communications between the customer and the technician. When a problem cannot be verified, the repair also cannot be verified. With a complete description of the problem, the technician can determine whether or not an effective diagnosis can be made.

Task A.5   **Establish relative importance of observed vehicle data.**

In any set or combination of vehicle symptoms, some are more important than others. For example, if a malfunction indicator lamp (MIL) is lit, one or more DTCs are almost certainly present. Reading and troubleshooting those DTCs would be the highest priority diagnostic tasks that relate to the set of symptoms for this car. If the symptoms also include a complaint about a rough idle or surging at cruising speeds, troubleshooting the DTCs and fixing their causes may also correct the other drivability problems.

Stored DTCs can have a greater impact on OBD II systems due to the interaction of components and the monitors that an OBD II computer runs. There is a hierarchy of diagnostic monitors in an OBD II system. The monitor-diagnostic routines run in a specific order. If one monitor fails to pass its system the other monitors will not run.

For example, a failed oxygen sensor will store a DTC for the sensor, but the computer also uses the sensor input to run the diagnostic monitor for the catalytic converter and possibly the EGR system. When the oxygen sensor code is set, the catalyst and EGR monitors are suspended, pending the correction of the oxygen sensor problem. Once the oxygen sensor is repaired, it is possible for another code to appear in different system after the diagnostic monitor that was "blocking" the other monitors passes it testing criteria. It may be necessary to run all monitors and recheck for codes once a component is replaced on an OBD II vehicle.

Task A.6   **Differentiate between powertrain mechanical and electrical/electronic problems, including variable valve timing (VVT) systems.**

Mechanical failures can mimic electronic problems and vice versa. Computerized engine controls depend on a mechanically sound engine. Many DTCs that appear in the computer system actually are caused by mechanical wear or failure in the engine. For example, if the MIL is lit and a DTC is present for a manifold absolute pressure (MAP) sensor problem, you should look for an intake vacuum leak besides testing the MAP sensor and its electrical circuit.

Task A.7   **Diagnose engine mechanical condition using an exhaust gas analyzer.**

Exhaust gas analyzers used in the repair field today will be either four- or five-gas analyzers. A four-gas analyzer measures carbon monoxide (CO), unburned hydrocarbons (HC), oxygen (O_2), and carbon dioxide (CO_2) in the exhaust. A five-gas analyzer measures the same four gases but also measures Oxides of nitrogen (NO_x).

Carbon monoxide is an odorless, colorless gas that is poisonous and must be properly vented away from the workplace. Carbon monoxide is formed in the combustion chamber when gasoline burns without a sufficient amount of oxygen present for complete combustion. When the air/fuel mixture is leaner than ideal (14.7:1), carbon monoxide levels will be low, usually less than 0.5 percent. When the air/fuel mixture is richer than ideal (14.7:1), carbon monoxide levels will rise. Carbon monoxide readings are used as a rich mixture indicator.

Hydrocarbons in the exhaust sample are from unburned fuel. Anything that inhibits proper combustion will increase hydrocarbon levels. Rich air/fuel mixtures, lean mixtures that create a misfire, ignition problems or compression problems will all increase hydrocarbon levels. Because hydrocarbons can increase when the air/fuel mixture is too rich or too lean, hydrocarbon levels cannot be used as a rich or lean mixture indicator.

Oxygen is found in the exhaust sample because 21 percent of the air drawn into the engine is oxygen and usually not all of the oxygen will be used during combustion. When the air/fuel mixture is richer than 14.7 to 1, the oxygen level will be low, generally less than 0.5 percent. When the air/fuel mixture is leaner than 14.7 to 1, oxygen levels will climb. If oxygen is greater than 1.2 percent, the mixture is either too lean, a misfire is present, or there is outside air being drawn into the analyzer.

If the vehicle has air injection, the system must be disabled prior to performing gas analysis, or false readings will result.

Carbon dioxide is a harmless byproduct of combustion. Carbon dioxide is used as a combustion efficiency indicator. The higher the carbon dioxide levels the better. A properly running late-model engine should produce carbon dioxide levels between 13.5 to 15.5 percent.

Oxides of nitrogen are produced in the combustion chamber when oxygen bonds with nitrogen during periods of operation when combustion chamber temperatures are high, generally above 2500°F. Oxides of nitrogen along with unburned hydrocarbons form photochemical smog in the presence of sunlight. The analyzer identifies the type and amount of excessive emissions, but the technician must determine the cause.

Excessive exhaust emissions can be caused by engine mechanical problems, as well as by problems in the fuel, ignition, and engine control systems. For example, low compression due to a burned or leaking valve will cause incomplete combustion. This results in increased HC and oxygen in the exhaust stream because the mixture does not burn in the combustion chamber. On an oxygen sensor equipped engine, the increased oxygen levels may cause the computer to believe the system is too lean and increase fuel delivery to all the cylinders in the engine, and this results in increased CO levels as well.

High hydrocarbon levels can be caused by anything that causes a misfire or incomplete combustion in the cylinder, as well as an over-fueling condition. Too much fuel being delivered to the engine causes high carbon monoxide readings. When the mixture is rich, the oxygen level should be low. If the engine has feedback control, it is possible to have high CO and O_2 readings if a cylinder is misfiring, as mentioned above. When the air/fuel mixture is ideal (14.7:1), the carbon dioxide level will be near its peak. A worn-out engine or inoperative catalytic converter will lower CO_2 levels.

Oxides of nitrogen are produced when the engine is under load and cylinder temperatures are high. Lean air/fuel mixtures and over-advanced ignition timing increase NO_x levels. The main job of controlling NO_x emissions lies with the exhaust gas recirculation (EGR) system. The catalytic converter reduces what remains in the exhaust stream to regulated levels. Because very little NO_x is produced when the engine in not loaded, most five-gas analyzers are portable so the vehicle can be driven while the exhaust gases are sampled. Proper operation of the EGR system is critical in controlling NO_x emissions because the catalytic converter will only reduce NO_x levels a small amount.

Task A.8   Diagnose drivability problems and emission failures caused by cooling system problems.

A low coolant level can cause the engine coolant temperature (ECT) sensor to warm up unevenly. An uneven, or erratic, signal from the ECT sensor will affect fuel metering and the air/fuel mixture. This in turn can lead to drivability and fuel economy problems, as well as increased CO and HC emissions.

An engine that is running too hot may tend to knock, which can cause the knock sensor to signal the PCM to retard ignition timing. Poor performance and fuel economy can result. An overheated engine also may suffer from incomplete combustion and a resulting increase in HC emissions.

If a cooling system problem prevents the engine from reaching normal operating temperature, the engine may run rich or may never reach closed-loop operation. Poor fuel mileage and sluggish performance may result.

If coolant enters the exhaust system through a combustion chamber leak, it will contaminate the O_2 sensor and eventually prevent it from working.

Task A.9   Diagnose drivability problems and emission failures caused by engine mechanical problems.

The adaptive learning capability of most late-model engine control computers can adjust to changes and wear in the engine, but no system can compensate for base engine mechanical problems that affect compression or airflow through the engine. A technician must verify engine integrity before attempting to diagnose problems with the engine control system. Vacuum tests, relative compression tests, cranking and running mechanical gauge compression tests and cylinder leakage tests may all need to be performed to verify engine mechanical integrity. Engine vacuum testing can determine the presence of cylinder sealing and airflow problems in the engine. Relative compression tests can help isolate individual cylinder problems. Cranking and running compression tests done with a gauge will pinpoint cylinder sealing or airflow problems. Cylinder leakage testing can pinpoint the exact cause of cylinder sealing problems such as a leaking intake or exhaust valves or head gasket failure.

An engine with a burned exhaust valve, resulting in low cylinder pressure, may show a continuously lean exhaust condition because of the additional oxygen in the exhaust due to incomplete combustion. This condition may be indicated by higher than normal O_2 readings on a four-gas exhaust analyzer or by a continuous lean-exhaust signal from the exhaust O_2 sensor.

A leaking or burned exhaust valve will not affect airflow, but it can cause incomplete combustion by not sealing the combustion chamber. A leaking intake valve also can reduce cylinder pressure and lead to incomplete combustion, and it can affect intake airflow.

Incomplete combustion resulting from low compression or a vacuum leak can let uncombined oxygen pass through the combustion chamber to the exhaust. This can cause continual lean-exhaust readings from the O_2 sensor (low voltage from a common zirconia O_2 sensor).

A worn camshaft lobe will reduce both lift and duration for its valve. Depending on whether the valve is an intake or exhaust valve, airflow into the cylinder or exhaust flow out will be affected. Cam lobe wear often is uneven among cylinders and results in rough running and uneven compression.

A timing belt or timing chain that has slipped will alter valve timing and generally affect all cylinders equally.

Leaking valves, broken or worn piston rings, and leaking head gaskets can all cause low compression. Depending on the amount of compression loss and the number of cylinders affected, the engine may run roughly and misfire.

Task A.10   Diagnose drivability problems and emission failures caused by problems or modifications in the transmission or final drive or by incorrect tire size.

Transmission shift speeds that are too high or too low can reduce vehicle performance and fuel economy. An automatic transmission that shifts at the wrong speed also can increase exhaust emissions during an I/M emission test drive cycle. If transmission shift points are altered by installing a "shift kit," emissions similarly may be increased.

The powertrain control module (PCM) or transmission control module (TCM) receives temperature sensor signals, speed sensor signals, and gear position signals from the automatic transmission, as well as input signals from engine sensors. The PCM or TCM uses combinations of these signals to determine when to lock up the torque converter and when to command upshifts and downshifts. Replacing the PCM (or the PROM inside the PCM) is sometimes done to fix drivability problems associated with torque converter lock-up. If the change originates with the carmaker, it will not adversely affect emissions or fuel economy. Aftermarket PROMs or PCM changes may adversely affect emissions and economy, however. The carmaker's TSBs are a good source of information about approved PCM modifications.

A torque converter that continuously locks and unlocks may create a symptom very similar to an engine misfire or surging condition. One good way to distinguish between an engine misfire or surge and a torque converter shudder is to drive the car until the symptom occurs and then lightly touch the brake pedal. Operating the brake switch commands the PCM or TCM to unlock the converter. If the symptom goes away, it was MOST likely related to torque converter lock-up, not to the engine.

Installing oversize wheels and tires on a vehicle without a corresponding correction for the vehicle speed sensor (VSS) can cause inaccurate speed signals to be sent to the PCM or TCM. Generally, oversize wheels and tires cause a VSS-speed signal that is slower than true vehicle speed. Any inaccurate speed signal can affect torque converter lock-up, shift points, and general engine performance. Wheels and tires other than the original equipment sizes should be selected from the car maker's approved optional sizes. The VSS signal often can be recalibrated for these optional tire sizes.

Task A.11   **Diagnose drivability problems and emission failures caused by exhaust system problems or modifications.**

When exhaust system components are changed in an attempt to improve performance, catalytic converters are often relocated or sometimes removed altogether. Original-equipment exhaust systems are designed for the best combination of exhaust flow and any slight back pressure necessary for EGR operation. Changing pipe diameters-usually from smaller to larger-and relocating mufflers and converters can upset original exhaust flow and back pressure requirements. More often than not, vehicle performance suffers.

Any change in EGR operation caused by exhaust system modifications can lead to increased NO_x emissions and engine pinging because EGR is a principal method used to control detonation.

Aftermarket exhaust systems must meet original-equipment specifications. In many cases, they also must be approved by government regulatory agencies. Aftermarket exhaust systems also must include exhaust oxygen sensors (O_2S), of original-equipment specifications and installed in original locations in the exhaust pipes or manifolds. Most importantly, a vehicle originally equipped with a catalytic converter may not have the converter removed.

A plugged catalytic converter or other restriction in the exhaust system will increase exhaust back pressure, which can lead to several drivability complaints. Most noticeably, the vehicle will lose power and perform sluggishly. Gas mileage may decrease, and the vehicle may fail an emissions test.

An exhaust leak upstream from the O_2 sensor may draw air into the exhaust stream and cause erroneous O_2 sensor readings. This, in turn, can lead to incorrect air/fuel mixture control by the PCM and may cause the system to drop out of closed loop into open-loop operation. Several different DTCs may occur.

Task A.12   **Determine root cause of failures.**

Task A.13   **Determine root cause of multiple component failures.**

Task A.14   **Determine root cause of repeated component failure.**

Tasks A.12, A.13, and A.14 are all interrelated, and they are equally related to similar tasks listed for the ignition, fuel, emission control, and electronic systems. These tasks are examples of step-by-step logic and critical thinking applied by you, the technician.

It is not enough just to replace a failed component without determining the cause of the failure. For example, if you replace an O_2 sensor and the new one fails in 5,000 to 10,000 miles, you did not fix the root cause of the failure. Perhaps a cooling system leak into a combustion chamber caused deposits to form on the O_2 sensor. Fixing the cooling system leak will fix the root cause of the O_2 sensor failure.

Several related components may fail from a single cause. An O_2 sensor and a catalytic converter could both suffer from the coolant leak described above. Depending on the severity of the leak and the length of time that it existed, it might be wise to check catalytic converter efficiency with a four-gas exhaust analyzer after replacing the second O_2 sensor.

The examples given illustrate the importance of finding the root cause of both multiple and repeated component failures. Other similar examples for specific problem solving tasks relating to the ignition, fuel, and emission systems are provided later in this overview.

B. Computerized Powertrain Controls Diagnosis (Including OBD II) (13 Questions)

Task B.1   **Inspect and test for missing, modified, inoperative, or tampered computerized powertrain control components.**

Changing, bypassing, or tampering with electronic engine control components violates Federal law and usually harms vehicle performance, fuel economy, and emission control. Part of troubleshooting an electronic control system is to inspect it for missing, damaged, or tampered components, just as you would inspect any other vehicle system.

Ensure that all vacuum hoses are connected correctly and free from damage. Also be sure that wiring harnesses are connected properly and free from damage and corroded connections.

An unattached connector on a wiring harness may be a clue that a device has been removed from the system. At this point, a component locator manual and a wiring diagram can be essential in determining if a part is missing.

If a vacuum solenoid is not controlling vacuum as it is supposed to, inspect the vacuum hoses and solenoid ports to see if they have been tampered with and intentionally plugged. A coolant sensor may be disconnected or have a resistor added to its signal wire to intentionally force the PCM to command a rich air/fuel ratio. This kind of tampering may be done in a misguided attempt to improve drivability, but it will harm fuel economy and emission control.

Similarly, inspect the throttle position (TP) sensor for obvious misadjustment or a bent tang. This kind of tampering is sometimes done to try to fix a rough or uneven idle, but it can throw the entire fuel-control program of the PCM out of range.

Task B.2   Locate relevant service information.

Gathering service information begins with understanding and verifying the driver's complaint or the vehicle symptoms. You also should find out if any service work has been done on the vehicle recently or if any aftermarket electronic accessories have been installed. If custom audio equipment or a cellular phone has been installed, these accessories may be the source of electromagnetic interference (EMI) that affects the control system. Additionally, a poor ground connection or power source for an accessory can cause problems with original electronic equipment on a vehicle. Learning about the service history and condition of a car is the act of getting firsthand information about the vehicle.

Service information for the powertrain control system can come from the same sources as other information about the car. Selecting the most relevant information can be the key to a fast and accurate diagnosis. Wiring diagrams may be particularly important for control system troubleshooting. Diagnostic flowcharts and "trouble trees" can be equally important.

Task B.3   Research system operation using technical information to determine diagnostic procedure.

Task B.3 is interrelated with several other similar tasks for this test. For accurate and efficient diagnosis, you must understand the operating details of a specific system. For example, some control systems have "default" or "limp-in" operating modes that let the system continue to operate backup values from its own memory in case of sensor failures. Other systems do not have such default operating modes. You must know the characteristics of the particular system you are testing.

As another example, some fuel-injection control systems have a "clear-flood" operating mode. In this mode, the PCM will reduce or cut off fuel flow during cranking if TPS voltage is above the midpoint of its range. Knowing whether or not a particular system has this characteristic can be an important part of diagnosing a no-start problem.

Additionally, adaptive-memory features vary from one system to another. This feature is the ability of a PCM to learn the operating characteristics of a particular vehicle and to compensate for age and engine wear. Resetting the adaptive memory—when applicable— is a key step in many service procedures.

Task B.4   Use appropriate diagnostic procedures based on available vehicle data and service information; determine if available information is adequate to proceed with effective diagnosis.

Troubleshooting a powertrain control system often requires several series of diagnostic procedures. You may find yourself obtaining a diagnostic flowchart and performing the test procedure, only to have to get more information and perform other tests. The overall diagnostic process can be repetitive. You may have enough information for the first rounds of diagnosis, but you may find that you do not have adequate information to confirm that you have identified the root cause of a problem.

To diagnose general powertrain control system problems, it is necessary to have a thorough understanding of system design principals and operating strategies. Systems that comply with OBD II diagnostic standards provide a minimum list of operating data parameters and have standardized DTCs. Accessing reliable repair and service information is essential in any diagnosis, including the parameters and criteria to set a DTC. Also, do not overlook reviewing any TSBs that could be related to the problem. This information can provide an excellent starting point for troubleshooting.

Note: Scan tool data includes data stream, diagnostic trouble codes, freeze frame data, system monitors, and readiness monitors.

Task B.5 — Determine current version of computerized power train control system software and updates; perform reprogramming procedures.

Due to changes made by the manufacturer of the vehicle's PCM operating system the technician must be able to determine the current version of the software in the power train control module and update by performing reprogramming procedures. Notification usually comes in the form of a technical service bulletin for emission changes, drivability changes, or false MIL operation. The updates are downloaded to a scan tool, then uploaded to the PCM.

Task B.6 — Research ODBII system operation to determine the enable criteria for setting and clearing diagnostic trouble codes (DTCs), and malfunction indicator lamp (MIL) operation.

The OBD II system runs system monitors to indicate all of the sensors within a part of the engine management system are working properly. In order for the PCM to run a monitor, certain conditions must be met. These conditions are called the enabling criteria; for example, the enabling criteria for the misfire monitor are valid signals from the MAP, MAF, ECT, VSS, and RPM. The RPM must be within a specific range, the ECT must be within a specific range, and the VSS also must be within a specific range. The clearing of trouble codes is accomplished with the use of a scan tool. Any diagnostic trouble code will eventually be erased from the PCM memory. The length of time a code will remain stored depends on the trouble code, but at least 40 warm-up cycles are required. The MIL is used to indicate problems in the components and systems monitored by OBD II.

Task B.7 — Interpret OBD II scan tool data stream, diagnostic trouble codes (DTCs), freeze frame data, system monitors, monitor readiness indicators, and trip and drive cycle information to determine system condition and verify repair effectiveness.

A scan tool is a test computer that communicates with the PCM and other onboard computers of a vehicle. The scan tool reads and displays diagnostic information provided by the PCM such as DTCs and system operating data, or parameters.

Data transmitted to the scan tool from the PCM include both digital and analog values. Digital parameters are often called switch parameters, or signals, and are either on or off, high or low, yes or no. An analog parameter provides a signal value with a specific minimum-to-maximum range. These kinds of data include analog voltage readings, speed signals, temperature readings, and frequency ranges.

Every item of data transmitted from the PCM to a scan tool has specific value or signal range described in the vehicle specifications. You must have access to, or knowledge of, these specifications and compare them to the scan tool readings to identify a system fault.

Scan tool readings that identify an open or a short circuit are among the easiest to recognize. If a resistive sensor displays a scan tool reading at or near the 5.0-volt reference voltage on which most such sensors operate, the sensor circuit to the PCM is open. If the scan tool voltage for such a sensor is at or near 0 volt, the circuit is probably grounded.

Scan tool readings of PCM input and output signals reflect values as processed by the PCM. In some systems, a sensor failure will cause the PCM to ignore the signal from the failed sensor and operate on backup values stored in its own memory. In this case, the PCM may transmit the backup values to the scan tool in place of the failed sensor signal. If any particular scan tool data reading does not make sense in relation to particular problem or symptom, you should test the system component directly with a voltmeter, ohmmeter, oscilloscope, frequency counter, or other test equipment.

OBD II diagnostic standards require a uniform library of diagnostic trouble codes (DTC) to be used by all carmakers. Additionally, all OBD II systems must transmit a basic list of 16 data items to a scan tool. Carmakers are free, however, to "enhance" the onboard diagnostic capabilities of their control systems. Many systems provide much more than the minimum OBD II diagnostic data requirements.

The OBD II system monitors virtually all emission control systems and components that can affect tailpipe or evaporative emissions. In most cases, malfunctions must be detected before emissions exceed 1.5 times the emission standards. If a system or component exceeds emission thresholds or fails to operate within a manufacturer's specifications, a DTC will be stored and the MIL will be illuminated within two driving cycles.

The OBD II system monitors for malfunctions either continuously, regardless of driving mode, or non-continuously, once per drive cycle during specific drive modes. A DTC is stored in the PCM when a malfunction is initially detected. In most cases the MIL is illuminated after two consecutive drive cycles with the malfunction present. Once the MIL is illuminated, three consecutive drive cycles without a malfunction detected are required to extinguish the MIL. The DTC is erased after 40 engine warm-up cycles once the MIL is extinguished.

In addition to specifying and standardizing much of the diagnostics and MIL operation, OBD II requires the use of standard communication links and messages, standardized DTCs and terminology. Examples of standard diagnostic information are freeze frame data and Inspection Maintenance (IM) Readiness Indicators.

Freeze frame data is stored in the PCM at the point the malfunction is initially detected. Freeze frame data consists of parameters such as engine rpm and load, state of fuel control, spark, and warm-up status. Freeze frame data is stored at the time the first malfunction is detected; however, previously stored conditions will be replaced if a fuel or misfire fault is detected. This data is accessible with the scan tool to assist in repairing the vehicle.

OBD II Inspection Maintenance (IM) Readiness indicators show whether all of the OBD II monitors have been completed.

Task B.8   Establish relative importance of displayed scan tool data.

PCM serial data transmitted to a scan tool may have a data list of 50, 60, or more items. Not all items of serial data are relevant to a single problem or symptom. Later sections of this overview discuss the most important serial data parameters for ignition, fuel, and emission system troubleshooting. For general powertrain control diagnosis, the following data parameters are among the most important:

- *System voltage*—The battery and charging system must provide a continuous regulated voltage of 12 to 14 volts. More importantly, the PCM must receive this voltage. Most serial data streams provide a reading of system voltage at the PCM. It must be within the normal range for the PCM and all system components to operate properly.

- *Engine speed, or rpm*—The engine rpm, or "tach" signal is the single most important input to the PCM. Without this signal, the PCM can't even tell if the engine is running. The tach signal tells the PCM whether the engine is cranking, idling, accelerating, decelerating, or cruising. The tach signal affects all ignition, fuel, emission, and transmission control operations of the PCM.

- *Vehicle speed, or mph*—Vehicle speed affects transmission control, ignition timing, fuel metering, and several emission-control subsystems. It also is an input signal for some antilock brake systems (ABS) and cruise control.

- *Temperature parameters*—The temperature of engine coolant, intake air, and transmission fluid affects fuel, ignition, and emission system operation, as well as transmission control.

- *Manifold pressure (or vacuum) and barometric pressure*—These air pressure parameters are primary inputs for fuel and ignition control and transmission shifting. Most control systems use air pressure measurements as a starting point for the PCM to calculate relative engine load.

For any given problem, only a few data parameters may be of primary importance for diagnosis. The ones listed are a few that affect several subsystems or overall system operation. Others are described in later sections of this overview. Many scan tool manufacturers' group related parameters together on their instrument displays or provide submenus of selected key data readings for specific subsystems. The most important parameters for overall system operation often are displayed at the top of a scan tool data display.

Control systems that comply with OBD II onboard diagnostic standards provide a minimum list of operating data parameters. These include most of the important items to evaluate for general system troubleshooting. Basic OBD II data parameters include: O_2 sensor voltage, open- and closed-loop indications, engine speed and temperature, barometric and manifold pressure, short-term and long-term fuel trim corrections, intake air temperature, spark advance, engine load, and vehicle speed.

Task B.9 Differentiate between electronic powertrain controls problems and mechanical problems.

Any given symptom or driver complaint can be caused by a problem in more than one system. A problem can appear to be based in one system when, in fact, it originates in another. For example, a longer than normal fuel injector pulse width and high long-term fuel trim correction factors are fuel system symptoms, but their cause can lie in several other engine subsystems. A mechanical problem, such as an intake vacuum leak, could be the root cause. On the other hand, an incorrect input signal from a barometric pressure sensor or manifold pressure sensor could cause these unusual fuel control symptoms.

Differentiating between mechanical problems and computer system problems is part of identifying and repairing the root cause of any symptom. As a diagnostic technician, you must look for abnormal test results and isolate the cause of a problem to one system rather than another.

Information is often the key to diagnosing or repairing late-model computer controlled systems. Diagnostic information will be available both in printed form and in computerized information retrieval systems in the shop or on the Internet. Additionally, many equipment companies provide databases of test procedures and specifications that are built into their test equipment that can be accessed by the technician while performing diagnostics with the equipment. This is an emerging and very useful technology that gives a technician information when it is needed. Technical service bulletins (TSBs) are usually high priority information for late-model systems and are included in all of the sources already mentioned. The technician must be aware of what service information is available and how to get the information needed. The best service information in the world is useless if the technician cannot readily retrieve the information.

Intermittent engine control problems will require diagnostic procedures tailored to the conditions present when the problem occurs. This service information needs to be gathered by careful questioning of the vehicle owner so that the technician can perform an effective diagnosis. Heating or cooling affected components could find temperature related faults. Performing wiggle tests and monitoring circuit values can find harness and connection problems. Problems such as these can go undetected if the technician is not provided with accurate service information.

Task B.10 Diagnose no-starting, hard starting, stalling, engine misfire, poor drivability, incorrect idle speed, poor idle, hesitation, surging, spark knock, power loss, poor mileage, illuminated MIL, and emissions problems caused by failures of computerized powertrain controls.

The proper operation of computerized powertrain controls is essential for the modern power plant to function correctly. A computer that loses power or ground will not function at all and will cause a no-start condition on any fuel-injected engine. All of the conditions listed in Task B.8 have the potential to be traced back to the computer and its input and output components.

While a failure in the computer-control system is a possible source of these conditions, it should not be considered the problem until basic mechanical and electrical tests have been performed. A manifold absolute pressure sensor (MAP) will not generate a correct signal if the engine is incapable of producing normal manifold vacuum. An out of calibration coolant sensor input can cause hard starting when the engine is both cold and hot, as well as altering fuel delivery enough to cause an emission test failure and poor mileage. A MAP sensor that is out of calibration can cause many different drivability problems on a computer-controlled engine because it is a primary sensor input. Fuel delivery, spark timing, or torque converter clutch lock-up can all be affected by the MAP input. Drivability concerns, such as hard starting, stalling, surging, spark knock, poor mileage, and excessive emissions, can all be traced to an incorrect MAP sensor signal. A problem with the throttle position sensor input may not only cause engine hesitation but may also affect transmission shift points or feel on electronically controlled transmissions. It is important to review the theory of operation section in the repair information for the vehicle to fully understand how each engine sensor input can affect the operation of the engine or transmission.

A failed output actuator, such as an EGR solenoid, may prevent the computer from properly controlling EGR gas flow. The computer may energize the EGR solenoid to allow vacuum to reach the EGR valve while other systems energize the solenoid to block vacuum to the valve. In the first case, if the EGR valve does not receive vacuum, the engine may produce spark knock. In the second case, if the computer cannot limit EGR flow, hesitation, surging, and power loss may be the result. In both cases, understanding system operation is the key to correct diagnosis. A failure inside the computer, such as an open driver transistor for a fuel injector or ignition coil, can cause engine misfire and power loss. Computers may also exhibit intermittent drivability concerns from problems such as solder joint connections that are temperature or vibration sensitive. These types of failures can be very challenging to diagnose and require employing advanced diagnostic procedures, such as circuit monitoring with such tools as labscopes and graphing multimeters. The table below lists a few common problems that can be associated to a specific symptom. The table is not intended to be all-inclusive but only to point out how a component failure can affect the vehicle.

Task B.11   Diagnose failures in the data communication bus network; determine needed repairs.

The data communication bus network is the phone line among all the different computers on the vehicle. Through the bus network the computers can share information, therefore eliminating wiring. For example, the power train control module can be hardwired to the throttle position sensor to monitor throttle operation. The transmission control module (TCM) needs information on throttle operation as well; however, the TCM can monitor throttle operation through the bus network rather than being hardwired to the sensor. Communication bus networks have gotten faster and faster with each evolution of the network. The latest version is the controller area network or CAN.

Task B.12   Diagnose failures in the anti-theft/immobilizer system; determine needed repairs.

The anti-theft/immobilizer system prevents the engine from starting without the key. The key has an immobilizer chip inside it that transmits a signal to the anti-theft module. The module determines if the vehicle is being started using the correct key. If so the module sends a signal to the power train control module allowing the vehicle to start. A malfunction in the anti-theft system can cause a no start.

Task B.13   Perform voltage-drop tests on power circuits and ground circuits.

Voltage-drop tests have always been important diagnostic methods for any electrical system. They have never been more important than they are for troubleshooting electronic control systems.

To perform a voltage-drop test, you must energize the circuit (turn on the ignition or close the appropriate switch) to ensure that current is flowing in the circuit. Without current flow, voltage cannot drop across all the loads in the circuit. You can measure voltage drop two ways:

- Connect your voltmeter directly across the circuit load, observe system polarity, and read the voltage drop directly.

- Connect your voltmeter negative lead to ground and connect the positive lead first to the positive side of the circuit load, then to the negative side. Subtract the second reading from the first to determine the voltage drop across that specific circuit load.

Use the following values for maximum voltage drops across circuit wiring, connections, and switches:

- 0.00 V across a connection

- 0.20 V across a length of wire or cable

- 0.30 V across a switch

- 0.10 V at ground or across a ground connection

Because electronic systems operate with low voltage and very low current, a clean ground connection with minimum voltage drop is essential. To measure voltage drop across the complete ground side of a circuit, connect your voltmeter negative lead directly to the negative battery terminal. Then probe with or connect the positive meter lead to ground connectors, terminals, and the battery ground cable itself. Total ground voltage drop should not exceed 0.10 volt.

Remember that the sum of all the voltage drops in a circuit must equal source voltage exactly. If the measured voltage drops of all the designed circuit loads are less than source voltage, some unwanted resistance exists in the circuit. This is usually a damaged connector or wiring, corrosion, or a poor ground connection.

Task B.14   Perform current flow tests on system circuits.

Performing current flow tests through system circuits will require a Digital Multimeter (DMM) and/or lab scope. Here is an example of a circuit and how you would perform the test and what you should see as readings: **Canister Purge Circuit.**

At ambient temperature above 68°F, fuel vapor pressure in the fuel tank and fuel system will rise when the vehicle is at rest. The canister purge circuit is designed to take advantage of this rise in pressure by channeling the fuel vapors into a "holding tank" called the Evaporative Canister. The vapors stored in the evaporative canister are typically brought to the engine via the canister purge valve. The flow rate through the canister purge valve is controlled by another solenoid called the vent valve.

To test the electrical circuit at the canister purge as well as the control of the canister purge, current testing can be used. To do this, perform the following steps:

- Remove the fuse for the canister purge solenoid.

- Connect an ammeter in place of the fuse.

- Using a scan tool, command the solenoid on.

- The reading on the ammeter should increase.

Note: A typical good value should show an amperage increase of 400–700 ma, with the solenoid off, amperage should decrease 400–700 ma.

Current testing is required whenever a computer that controls an output device is being replaced. Output devices include items such as mixture control solenoids, EGR, canister purge or shift solenoids, fuel injectors or the control circuit coil in a relay. A low resistance solenoid or coil will increase current flow and possibly destroy a driver transistor in the computer. If the computer is replaced without identifying the defective solenoid or coil, the replacement computer will suffer the same failure. Current testing output devices should be done with a DMM set to the amps scale and connected with the red lead to each output terminal in the computers connector and the black lead to ground with the ignition switch in the Run position. The reading should be taken over a five-minute time span to allow the device to heat up, and the value should not exceed manufacturer's specifications, usually less than 750 milliamps. Low resistance solenoids, such as pulse width modulated (PWM) transmission solenoids or fuel injectors that use current limiting drivers (Peak and Hold), cannot be tested in the above manner due to excessive current flow. A labscope and current probe or resistance testing with an ohmmeter will be required.

Task B.15   Perform continuity/resistance tests on system circuits and components.

Continuity/resistance tests have always been important diagnostic methods for any electrical system. They have never been more important than they are for troubleshooting electronic control systems. To perform a continuity/resistance test, you must de-energize the circuit (turn off the ignition or open the appropriate switch) to ensure that no current is flowing in the circuit. With current flow, resistance cannot be measured and damage to the digital multimeter can occur. Some tests require the component be disconnected completely from the circuit. Be sure the right scale is selected on the multimeter if it is not a self ranging meter. Connect the leads across the two thermals and read the resistance, then compare it with manufacturers' specifications. A reading of OL does not mean zero resistance; it means the resistance of the circuit being measured is out of the meter's limit to measure. A reading of 0 does not mean nothing; it means the circuit being measured had 0 Ohms of resistance.

Task B.16   **Test input sensor/sensor circuit using scan tool data and/or waveform analysis.**

Signal analysis using a labscope or graphing multimeter is becoming one of the most important skills a drivability technician can possess today. Knowing the difference between good and bad signals requires training so that the technician understands circuit operation and how to use the equipment effectively. Technicians must also understand how to test known good system components. Many of today's automotive digital storage oscilloscopes (DSOs) allow a technician to save screens and store them in a computer software program for retrieval.

These programs provide the means to build a database of good and bad waveforms and can also print out waveforms for the customer in order to show repair verification. Some equipment also has known good waveforms stored in memory in the tool for comparison to the signal under test. Voltage testing an input signal alone does not provide the level of detail necessary to diagnose problems in the complex engine control systems used today. The DSO can pinpoint signal dropouts and glitches far better than the best DMMs can.

Task B.17   **Test output actuator/output circuit using scan tool, scan tool data, and/or waveform analysis.**

Testing procedures for output actuators using scan tools and waveform analysis can be enhanced if the system supports bi-directional testing. Using the scan tools bi-directional controls, many computer outputs can be turned on and off by the technician and measurements taken with a DSO. Even without bi-directional control the DSO can confirm output actuator operation based on computer commands monitored through scan tool data. Checking signals right at the PCM connector is generally the best method to determine if a circuit is operating properly. For instance, the DSO can show a canister purge or TCC solenoid turning on and off as commanded during a test drive while the scan data parameter for that circuit is also checked with the scan tool.

Task B.18   **Confirm the accuracy of observed scan tool data by directly measuring a system, circuit, or component for actual value.**

Using direct measurement to confirm accuracy of displayed scan tool data generally involves performing tests to determine calibration errors. An infrared temperature measuring tool may be used to check cylinder head temperature where the coolant sensor is mounted while also viewing the sensor scan parameter on a scan tool. Too large a discrepancy between the two readings means the coolant sensor is out of calibration. A similar test is performed on a manifold pressure sensor using a handheld vacuum pump to supply vacuum and a DMM to measure the sensor signal voltage. Two readings are taken at different vacuum levels and one reading subtracted from the other to determine proper sensor calibration. Performing this type of testing is crucial to eliminate incorrect diagnosis of drivability problems and unnecessary parts replacement.

Task B.19   **Test and confirm operation of electrical/electronic circuits not displayed in scan tool data.**

Voltage-drop testing discussed under task B.13 is the most common test principle applied to electrical and electronic components for which serial data is not available. You also can use actuator test mode (ATM) tests or output state checks to analyze the operation of output devices for which serial data is not available.

Verifying the B+ + voltage supply to the PCM and to other system components is a basic part of troubleshooting. The PCM and output devices for which the PCM provides ground control receive battery voltage from the vehicle electrical system. Many system sensors receive a 5.0-volt reference voltage from the PCM.

The best way to check any supply (B+) voltage is with a high-impedance digital voltmeter. A simple probe light should not be used because it only indicates that some voltage is present; it does not indicate the actual voltage level. Also, a probe light may draw excessive current through a circuit branch and damage electronic components.

Along with verifying the voltage supply to system devices, you must verify proper ground connections. You can identify an open ground connection by simple inspection or by measuring open-circuit voltage at a point in a circuit where you would expect a voltage drop. A high-resistance ground is best identified by voltage-drop measurement.

You can use an ammeter to verify and measure current in an operating circuit; and you can use an ammeter or a voltmeter, or both, to check the operation of solenoids, relays, and motors.

Task B.20   Determine root cause of failures.

Task B.21   Determine root cause of multiple component failures.

Task B.22   Determine root cause of repeated component failures.

Tasks B.20, B.21, and B.22 are related to each other and to similar tasks listed for other subsystems. An example of these principles applied to the control system in general is the case of PCM replacement because of damaged output driver transistors. If you replace a PCM because of transistor failure, you must pinpoint the cause of the transistor damage or the replacement unit probably would be damaged also.

PCM driver transistors are often damaged by short circuits in output solenoids, relays, motors, and fuel injectors. A short circuit in one of these devices can increase current flow through its PCM transistor and lead to failure. Identifying and replacing the defective output device isolates the root cause of the failure and avoids repeated failures.

A case history exists of a PCM that was replaced because of transistor damage caused by a shorted mixture control solenoid in the carburetor. This short circuit in the solenoid not only took out the PCM, it allowed the engine to run overly rich and eventually melt down the catalytic converter. Several hundred dollars worth of PCM and catalytic converter were destroyed because a $20 solenoid failed.

Task B.23   Verify effectiveness of repairs

After a repair is made to a vehicle the last step in the repair is to verify that the repair fixed the original complaint. This is usually done by test driving the vehicle in an attempt to reproduce the complaint. The original problem may require certain enabling criteria and/or a particular drive cycle in order for the PCM to run a system monitor test on a particular system.

C. Ignition System Diagnosis (7 Questions)

Task C.1   Inspect and test for missing, modified, inoperative, or tampered components.

The distributor is one of the ignition components most susceptible to tampering or modification. If the original distributor has been replaced by an aftermarket component, the replacement must meet original-equipment specifications. In many cases, a replacement distributor also must be approved by a government regulatory agency. An ignition distributor that is not approved and that does not meet original specifications can harm engine performance and increase emissions.

Spark timing control devices used on older computer-controlled engine systems and on non-computerized vehicles also were often modified or tampered with. Control of distributor vacuum advance mechanisms was a common way to reduce both HC and NO_x emissions. Solenoid valves were often used to cut off vacuum advance until a vehicle reached cruising speed. This was a common way to reduce HC emissions. Cutting off vacuum advance entirely until an engine reached very high temperature also was a common NO_x control method.

Any spark timing devices installed as original-equipment emission controls must be in place and operational. Component locator and identification manuals and test procedures will help you determine if such devices are present and working properly.

Task C.2   Locate relevant service information.

Service information for the ignition system can come from the same sources as other information about the car. Among the information or service data you will need for the ignition system are:

• Base timing setting and adjustment speed

- Spark advance data, or "curves"
- Correct spark plug part numbers, including heat range and resistor requirements
- Spark plug gap and installation torque
- Ignition cable (spark plug wire) resistance

For electronic spark timing, or a computer-controlled spark timing system, you must also know how to take the computer out of its spark advance control function to check and adjust base timing. This is often done by disconnecting a specified timing connector or installing a jumper wire.

Computer-controlled spark timing systems also use various sensors to indicate crankshaft position and to identify cylinders by number. You must know the number and kinds of ignition control sensors on a particular system for accurate troubleshooting.

As with other control system components and subsystems, technical service bulletins (TSB) are important sources of ignition system service specifications and diagnostic instructions.

Task C.3   Research system operation using technical information to determine diagnostic procedure.

A technician should develop a diagnostic routine based on a thorough understanding of system operation. Prior to starting a diagnosis the technician should research system operation in the service material that is available to the technician. Most computer-based information systems, as well as manufacturer's printed manuals, begin a service section with an overview of the system operation so that the technician is familiar with the normal operation. This material should not be overlooked. Service information will usually include symptom-based diagnostic procedures to help guide the technician in performing the needed tests to diagnose a complaint. Knowing what type of primary triggering device is used and whether the signal goes to an ignition control module (ICM) or directly to the PCM is necessary information when determining a diagnostic approach to a no-start condition.

Task C.4   Use appropriate diagnostic procedures based on available vehicle data and service information; determine if available information is adequate to proceed with effective diagnosis.

Some of the simplest ignition system checks are resistance tests of spark plug cables and ignition coils with an ohmmeter. If your ohmmeter indicates infinite resistance, you know that an open circuit exists in the component. For complete testing, however, you must know the manufacturer's resistance specifications. Without these specifications you can't pinpoint a low-resistance or high-resistance condition that may be leading to a short-circuit or open-circuit problem.

Service information that describes the type of engine sensors and their voltage waveform signals will help you to choose diagnostic procedures for those specific sensors. Without test procedures and waveform information, you may be able to determine if voltage is present or absent in a circuit, but you cannot tell if the signal is being received and understood by the ignition module or the PCM.

Task C.5   Establish relative importance of displayed scan tool data.

The following data parameters are among the most important for ignition system troubleshooting and to determine ignition effects on other subsystems:

- *Engine speed, or rpm*—The engine rpm, or "tach" signal tells the PCM whether the engine is cranking, idling, accelerating, decelerating, or cruising. The tach signal affects all ignition control operations of the PCM.

- *Crankshaft position (CKP) signal*—The CKP signal usually provides the engine rpm information to the PCM, and it indicates the angular position of the crankshaft and provides cylinder identification for spark advance control.

- *Spark advance*—Many systems provide direct data stream readings of spark advance.

- *Manifold pressure (or vacuum) and barometric pressure*—These air pressure parameters are primary inputs for ignition control. The MAP sensor in a late-model engine takes the place of the vacuum advance diaphragm on an older distributor.

- *Vehicle speed*—Vehicle speed affects ignition timing and several other control subsystems.

- *Temperature parameters*—The temperature of engine coolant and intake air affects ignition operation.

- *Throttle position*—The TPS signal has an immediate effect on electronic spark timing as the PCM changes spark advance for acceleration and deceleration, as well as at idle.

Control systems that comply with OBD II onboard diagnostic standards must provide data parameters for engine speed, spark advance, manifold pressure, throttle position, and coolant temperature as part of the minimum list of operating data parameters.

Task C.6   Differentiate between ignition electrical/electronic and ignition mechanical problems.

Electrical problems can mimic mechanical problems and vice versa. In an ignition system, intermittent problems in circuits between the ignition control module (ICM) and the PCM can cause erratic spark timing and engine misfire. The same performance symptoms also can result from mechanical damage to ignition sensors like the CKP sensor. To distinguish between an electrical or mechanical fault as the root cause of a problem, you must inspect system components for obvious damage and perform electrical tests to identify circuit problems.

Task C.7   Diagnose no-starting, hard-starting, stalling, engine misfire, poor drivability, spark knock, power loss, poor mileage, illuminated MIL, and emission problems on vehicles equipped with *distributorless electronic ignition* (EI) systems; determine needed repairs.

Some tests for distributorless ignition systems are unique to these kinds of systems; others are basic checks that apply to all ignition systems. Battery voltage must be supplied to one side of the primary windings for the coils and switching control must be provided on the ground side. Absence of battery voltage or primary switching can cause hard starting or a no-start condition. Similarly, open circuits in spark plug cables and ignition coils can cause the same problems.

When misfiring, detonation, poor performance, and increased emissions are related to the ignition system, they often result from spark timing problems. In a distributorless EI system, timing is controlled by the PCM. Timing problems can result from erratic sensor input signals and circuit problems between the PCM and the ICM. Some traditional causes of misfiring include breakdown of the ignition coils under load, open or shorted spark plug cables, and worn or damaged spark plugs.

The knock sensor input is an important feature of most computer-controlled ignition systems. When this sensor detects engine detonation, it signals the PCM to retard ignition timing (or decrease the advance) and increase EGR flow until the detonation stops.

Task C.8   Diagnose no-starting, hard-starting, stalling, engine misfire, poor drivability, spark knock, power loss, poor mileage, illuminated MIL, and emission problems on vehicles equipped with *distributor ignition* (DI) systems; determine needed repairs.

Most of the operations summarized under Task C.7 above apply to distributor-type ignition systems as well as to distributorless systems. With a distributor system, however, you are working with only one ignition coil instead of several. A distributor system also has additional mechanical features to check during troubleshooting.

If the distributor contains the ignition timing sensor (which functions as a CKP sensor), you must check the sensor connections and the sensor mounting. Most distributor sensors are magnetic pickup coils, Hall-effect sensors, or optical sensors. Hall effect and optical sensors require a voltage supply connection, as well as a signal connection, and all sensors require a secure ground. Problems with distributor pickups or sensors can cause hard starting, failure to start, misfiring, and other performance and emission problems.

Mechanical wear in the distributor shaft and its bushings can cause irregular firing from one cylinder to another and resultant engine misfiring. Such mechanical wear often can be spotted by viewing the ignition parade pattern on an oscilloscope.

Distributor caps and rotors can develop short circuit paths to ground for the secondary voltage. These paths are often called "carbon tracking" because of the conductive deposits that cause the short circuit. Carbon tracking or simple cracks in a cap or rotor can cause the secondary voltage to short to ground and prevent the spark plug from firing.

Task C.9   Test for ignition system failures under various engine load conditions.

The ignition system "available voltage" is the maximum secondary voltage that the system can deliver. The "required voltage" is the secondary voltage necessary under any condition to fire the spark plug. The required voltage varies with changes in engine speed and load, but available voltage must always be greater than required voltage. The difference between the two is the "voltage reserve."

Testing for ignition faults under engine load is the process of determining available voltage, required voltage, and voltage reserve. Problems that can lower the available voltage include:

- Less than the required primary voltage supplied to the coil primary windings.

- High resistance anywhere in the coil primary circuit.

- Shorted primary or secondary windings in the coil.

- Open or high resistance primary or secondary windings in the coil.

Some problems that can raise the required voltage include:

- Open circuits or excessive resistance in the distributor cap or rotor or spark plug wires.

- Worn spark plugs or wide plug gaps.

- Lean air/fuel mixtures.

- Excessive carbon deposits in combustion chambers that increase cylinder pressure and temperature.

Any of these problems can cause ignition misfire under load if the required voltage is greater than the available voltage. Problems that create a lower resistance path to ground in the secondary circuit will also cause misfire. Spark plug wire insulation leakage, torn spark plug boots, or cracked spark plug insulators will all allow secondary voltage to leak to ground rather than jumping the spark plug gap. Tests such as power braking (brake torque) the engine or cranking KV tests will stress the secondary circuit and identify these types of faults.

Task C.10   Test ignition system component operation using waveform analysis.

The use of the lab scope gives the technician a graphed visual picture of the activity of the ignition system component. The technician should have an understanding of the equipment being used and understand the different types of patterns he or she will encounter while testing this system. The technician will be testing primary, secondary, coils, wires, crank sensors, cam sensors, and spark plugs.

Task C.11   Confirm base ignition timing and/or spark timing control.

Checking ignition timing and spark advance used to be as simple as aiming a timing light at the timing marks on the engine crankshaft damper. If timing needed adjustment, you loosened the distributor hold-down bolt and turned the distributor to align the specified marks with the engine running at idle. Accelerating the engine and watching the timing advance on the crankshaft damper checked the operation of centrifugal advance. Applying vacuum to the distributor vacuum diaphragm checked the vacuum advance operation.

The simplest distributor-type electronic ignitions have centrifugal and vacuum advance mechanisms. You can check base timing and spark advance as you would on an older breaker-point system. Electronic spark timing (or computer-controlled spark advance) eliminated centrifugal and vacuum advance, but distributor systems still require base timing test and adjustment. To check base timing on these later systems, you must take the computer out of the timing control loop by disconnecting or jumpering a specified connector. Spark advance with computer-controlled spark timing generally is not tested. Distributorless electronic ignitions do not require a base timing test or adjustment.

If base timing is retarded and spark advance is less than required, fuel economy and performance can suffer. CO emissions also may increase. If base timing is overly advanced and spark advance is excessive, the engine may detonate and misfire. HC and NO_x emissions also can increase.

Task C.12   Determine root cause of failures.

Task C.13   Determine root cause of multiple component failures.

Task C.14   Determine root cause of repeated component failures.

Tasks C.12, C.13, and C.14 are related to each other and to similar tasks listed for other subsystems. An example of these principles applied to the ignition system is the case of an ignition module that fails repeatedly because of high temperature. High temperature can be caused by excessive current draw due to shorted ignition coil primary windings. High temperature faults also can result from high underhood temperatures. Engine overheating, missing heat shields, or high ambient temperatures can be the root cause of repeated module failures.

Spark plugs that foul after very few miles of operation are a classic example of a root cause located in another engine system. In this case, mechanical failures like broken rings or worn valve guides can cause the plug fouling, and new plugs will continue to foul until the root cause is repaired.

D. Fuel Systems and Air Induction Systems Diagnosis (7 Questions)

Task D.1   Inspect and test for missing, modified, inoperative, or tampered components.

The carburetor was the fuel system component most susceptible to tampering. Vehicle owners couldn't keep their hands off idle speed and mixture adjustments. Although easily adjustable carburetors are history, many late-model units have been misadjusted in attempts at a fast fix for idle problems.

Inspect carburetors for missing caps or plugs on mixture adjustment screws. Also look for attempts to adjust idle speed by other than original-equipment means. Inspect chokes for missing or misadjusted linkage. Check connectors to electric chokes for damage and signs of tampering.

If the original carburetor has been replaced by an aftermarket unit, the replacement must meet original-equipment specifications. In many cases, a replacement carburetor also must be approved by a government agency. A carburetor that is not approved and that does not meet original specifications can harm engine performance and increase emissions.

Air induction systems must meet original-equipment specifications and have all components that were originally installed on the car. Again, any aftermarket air cleaners or other air intake parts must meet original specifications and, in many cases, must be approved by a government agency. Modified air intake systems can increase HC and CO emissions, particularly at very cold and very hot temperature extremes.

Fuel-injection systems are less susceptible to tampering than carburetors. Some items to check for include aftermarket performance parts such as adjustable fuel-pressure regulators, high flow fuel injectors, intake manifolds or modified airflow sensors. All of the air intake components must be in place and functional. For both fuel-injected and carbureted engines, fuel pumps, filters, and delivery systems also must meet original-equipment specifications. Fuel delivery systems are interconnected with fuel evaporation emission systems. In some cases, modifications to the fuel delivery system can increase evaporative emissions.

Task D.2   Locate relevant service information.

Service information for the fuel and air intake systems comes from the same sources as other information about the car. Among the information or service data you will need for the fuel system are:

• Idle speed and mixture adjustment specifications and procedures for carbureted engines

- Air and fuel filter part numbers

- Fuel pressure specifications and test procedures, particularly for fuel-injected engines but for carbureted engines as well

For electronic idle speed control, you must also know how to make the minimum air-rate, or slow-idle, adjustments. For any specific fuel-injection system, you must know (or look up) the method used to measure intake air mass. That may be speed-density (manifold pressure and rpm), air velocity (vane airflow sensor), or mass airflow (air molecular mass, or weight). You also must know the number and kind of O_2 sensors used on the engine. As with other engine control components and subsystems, technical service bulletins (TSB) are important sources of fuel system service information.

Task D.3   Research system operation using technical information to determine diagnostic procedure.

For the technician diagnosing problems with fuel systems or air induction systems, the proper starting point must be with a thorough understanding of system operation. Using printed or electronic service information, the technician should research how the system is configured and what testing procedures the manufacturer outlines. A search of technical service bulletins should also be performed before detailed testing is started. Determining specifications (such as fuel pressure and volume), component locations (such as fuel pump relays or inertia switches), and air induction system components (such as intake manifold tuning valves or runner controls) will speed testing and diagnosis.

Task D.4   Evaluate the relationships between fuel trim values, oxygen sensor readings, and other sensor data to determine fuel system control performance.

The following data parameters are important for fuel and air induction system troubleshooting and for determining the effects of these systems on other subsystems. Whatever the oxygen sensor does, the fuel trim corrects. For example, if the oxygen sensor is seeing a lean air/fuel ratio, the short-term fuel trim will begin adding fuel; as the short-term fuel trim adds fuel, the long-term fuel trim will add fuel in order to lower short-term fuel trim.

O_2 Sensor—The exhaust oxygen sensor (O_2S) is the feedback signal that the PCM uses to control air/fuel ratios, and also is the primary input that the PCM uses to determine open- or closed-loop fuel control operation. When the O_2 sensor measures excess oxygen in the exhaust, the fuel mixture is too lean and the PCM will respond by making the mixture slightly richer. When the O_2 sensor measures less than normal oxygen in the exhaust, the fuel mixture is too rich and the PCM will respond by making the mixture slightly leaner.

Long-term and short-term fuel trim correction—Many fuel-injection systems provide data parameters that indicate the long-term trends and short-term actions to correct a fuel mixture in either the lean or the rich direction. OBD II onboard diagnostic standards require that these values be given as percentages.

Task D.5   Use appropriate diagnostic procedures based on available vehicle data and service information; determine if available information is adequate to proceed with effective diagnosis.

One of the most basic fuel system diagnostic sequences is to determine if the engine control system is operating in open loop or closed loop. The voltage signal from the O_2 signal is the primary indication of loop status, both to the vehicle PCM and to you, the technician.

After establishing whether the system is in open or closed loop, you can evaluate scan tool data and four- or five-exhaust analyzer readings to determine if the PCM is providing a rich or a lean correction and the extent of any such correction. The scan tool data items summarized under Task D.6 are the primary indicators of both loop status and rich or lean operation. Determining these basic conditions of control system operation is the first step in troubleshooting the fuel subsystem and other areas of the engine control system, as well.

Task D.6   Establish relative importance of displayed scan tool data.

The following data parameters are the most important for fuel and air induction system troubleshooting and for determining the effects of these systems on other subsystems:

- *Engine speed, or rpm*—The engine rpm, or "tach" signal tells the PCM whether the engine is cranking, idling, accelerating, decelerating, or cruising. The tach signal affects all ignition control operations of the PCM.

- O_2 Sensor—The exhaust oxygen sensor (O_2S) is the feedback signal that the PCM uses to control air/fuel ratios, and also is the primary input that the PCM uses to determine open- or closed-loop fuel control operation. When the O_2 sensor measures excess oxygen in the exhaust, the fuel mixture is too lean and the PCM will respond by making the mixture slightly richer. When the O_2 sensor measures less than normal oxygen in the exhaust, the fuel mixture is too rich and the PCM will respond by making the mixture slightly leaner.

- *Loop status (open or closed)*—This is an internal PCM parameter that indicates the status of fuel mixture control.

- *Intake air mass*—All fuel-control systems that transmit serial data provide some indication of intake air measurement. Air measurement may be based on (1) speed density (manifold pressure and rpm), (2) air velocity (vane airflow sensor), or (3) mass airflow (air molecular mass, or weight). Data readings may be in voltage, frequency, or grams per minute of airflow.

- *Long-term and short-term fuel trim correction*—Many fuel-injection systems provide data parameters that indicate the long-term trends and short-term actions to correct a fuel mixture in either the lean or the rich direction. OBD II onboard diagnostic standards require that these values be given as percentages.

- *Idle speed control operation*—The engine rpm signal indicates actual idle speed. Most systems also provide data on desired idle speed and on the operation of idle air control (IAC) valves or throttle solenoids and motors.

- *Temperature parameters*—Engine coolant temperature (ECT) and intake air temperature (IAT) are two primary input factors that the PCM uses to determine the air/fuel ratio.

- *Throttle position*—The TPS signal has an immediate effect on fuel control as the PCM changes air/fuel ratio for acceleration, deceleration, cruising, and idle.

- *Fuel pressure*-OBD II onboard diagnostic standards require a data readout of fuel pressure.

- *Mixture control solenoid operation for a feedback carburetor*—This data item may appear as a direct duty-cycle percentage or as a dwell measurement.

- *Vehicle speed*—Vehicle speed also affects fuel control.

Control systems that comply with OBD II onboard diagnostic standards must provide data parameters for engine speed, loop status, fuel trim correction, fuel pressure, intake air mass, throttle position, coolant and air temperature, and O_2 sensor readings as part of the minimum list of operating data parameters.

Task D.7   Differentiate between fuel system mechanical and fuel system electrical/electronic problems.

A faulty fuel pressure regulator is an example of a mechanical fuel system problem. The fuel pump relay and fuel pump are examples of electrical fuel system components. When troubleshooting a drivability complaint, a technician must be able to test all components of a fuel system, both mechanical and electrical, to determine the needed repair.

Task D.8   Differentiate between air induction system mechanical and air induction system electrical/electronic problems, including electronic throttle actuator controls (TAC) systems.

A vacuum leak is the most common example of a mechanical problem with the air induction system that will cause abnormal serial data readings and electronic control operation. Just as too much air entering the intake system through a leak will cause problems, too little leaving through the exhaust will cause its own set of problems. A restricted exhaust is a mechanical problem that will affect data for manifold pressure, fuel-injection pulse width, and O_2 sensor signals. On the air induction side of the intake system, an air leak in the intake duct or plenum may or may not affect fuel control. If the fuel-injection system is the speed density type that uses manifold pressure to calculate intake air density, an air leak in the intake duct will have no effect. The extra air entering the manifold is sensed and measured by the MAP sensor and accounted for in the air/fuel ratio. If, on the other hand, the fuel-injection system uses a mass airflow (MAF) sensor, and the intake air leak is downstream from the MAF sensor, the extra airflow will not be measured. This kind of intake air leak can cause abnormal lean mixtures and may set a DTC for the fuel system.

Task D.9   Diagnose hot or cold no-starting, hard starting, stalling, engine misfire, poor drivability, incorrect idle speed, poor idle, flooding, hesitation, surging, spark knock, power loss, poor mileage, dieseling, illuminated MIL and emission problems on vehicles equipped with *fuel-injection* fuel systems; determine needed action.

All of the traditional fuel injection problems that existed before computer control still exist today. A major fuel-delivery problem for fuel-injection systems is incorrect fuel pressure, which will be discussed for Task D.10. Incorrect fuel pressure can cause or contribute to many of the symptoms listed here.

Another major fuel injection problem that can cause many of these symptoms is improper fuel flow through the injectors. Restricted injectors can cause problems in either hot or cold starting and all phases of operation. Restricted injectors tend to cause lean mixtures and emission problems related to lean air/fuel ratios. Leaking injectors can contribute to many of the same problems as restricted injectors. Leaking injectors, however, tend to cause rich mixtures and the gas mileage and emission problems related to rich air/fuel ratios.

Vehicle manufacturers' service manuals often contain diagnostic flowcharts and procedures for each of these symptoms. This is another example of the importance of having complete and accurate diagnostic information.

Task D.10   Verify fuel quality, fuel system pressure, and fuel system volume.

Verifying fuel quality is a step that should not be overlooked when diagnosing performance complaints related to the fuel system. Water in the fuel, stale fuel, or excessive alcohol content will all have an adverse effect on engine performance. Commercially available test kits can be obtained that allow measurements to be made concerning fuel quality, such as alcohol content and volatility.

To test fuel pressure for a carbureted engine, disconnect the fuel line at the carburetor and install a gauge with a T-fitting. Crank or start the engine and monitor the pressure shown on the gauge: Refer to manufacturer's specifications for the vehicle being tested. To check fuel pump volume, disconnect the fuel line at the carburetor and direct fuel flow into an approved safety container. The pump should deliver 1 pint of fuel in 10 to 15 seconds.

Most port fuel-injection systems have a pressure test port on the fuel rail. Many throttle-body systems also have a test port. If the system has such a test port, it normally is fitted with a Schrader valve. Simply connect your test gauge to the port and crank or start the engine to check pressure. If the fuel-injection system does not have a test port, install a gauge with a T-fitting.

As a broad, general rule, port fuel-injection systems operate at higher pressures than throttle-body systems. Pressures for PFI systems can range from 28 to 65 psi; TBI system pressures range from 9 to 30 psi. These are only general limits, however; you must have the carmaker's exact specifications for accurate testing. If fuel-pressure values are used in a question on the test, the specifications required to make a diagnosis are given in the question.

Fuel pressure regulated by the vacuum regulator with the engine idling will be 8 to 10 psi lower than with no vacuum applied to the regulator. The vacuum regulator maintains a constant pressure drop across the fuel injector in response to changing intake manifold pressure present at the injector's tip. When intake manifold pressure rises, fuel pressure rises to keep fuel delivery constant. A fuel pump should produce pressure that is higher than the maximum regulated pressure. Momentarily clamping off the fuel return line or deadheading the pump can measure this pressure. Almost all fuel-injection systems have fuel pump check valves and regulators that close to maintain system pressure at rest when the pump stops. This rest pressure should be checked. Any rapid drop in rest pressure means a problem and should be diagnosed by clamping off the feed and return lines alternately to isolate where the leak is from. If the pressure remains constant with the feed line clamped, a leaking fuel pump check valve is indicated. Steady pressure with the return line clamped means a leaking pressure regulator, and, if leak-down occurs with both lines clamped, the injectors are leaking.

Task D.11   Evaluate fuel injector and fuel pump performance (mechanical and electrical operation).

Faulty fuel injectors can cause uneven idle speeds and surging or hesitation at cruising speeds. Emission test failures for high CO can indicate a rich condition caused by leaking injectors. Similarly, high HC readings can be caused by a lean misfire due to plugged injectors. High, short spark lines on an ignition oscilloscope secondary waveform also can indicate a lean mixture caused by restricted injectors.

On the car, testing of fuel injectors for flow rate is done with a pressure drop test. Each injector is operated one at a time with a special fixed rate pulse tool, while the pressure drop in the fuel rail is monitored with a gauge. A lower than average pressure drop indicates a lean or restricted injector while a greater than average pressure drop indicates a rich or leaking injector. More accurate car injector test equipment can measure fuel flow in milliliters and also check spray patterns. This type of equipment will also clean the injectors during testing and injector sets can be flow matched for optimal performance.

Injectors can be tested electronically with a digital storage oscilloscope. Both voltage and current waveforms can indicate problems with the circuit or the injector itself. Low inductive spikes seen on a voltage waveform can indicate a shorted injector coil. Fuel injector feed voltage and ground levels can be measured on a voltage waveform. A low amps current probe does an even better job of indicating shorted or high resistance injector coils and can also display injector pintle opening events in the waveform. This test can pick out sticking injectors that may otherwise be difficult to find. A low amps probe tests the injector coil under full voltage, dynamic conditions and is thus much more accurate than testing injector resistance with an ohmmeter.

Many late-model engine control systems for sequential fuel injection have built-in self-tests in which the vehicle PCM turns off each injector in sequence. The PCM then records the rpm drop and indicates on the data stream which injectors may not have provided a sufficient speed drop. This test is similar to the cylinder balance test provided on many engine analyzers, in which the secondary ignition is disabled cylinder by cylinder. You must remember, however, that the fuel-injection balance test only indicates that a cylinder is not delivering full power. The injector or other components may be the cause.

Evaluating fuel pump performance mechanical operation is done with pressure and volume tests listed in Task D.10. Even a mechanically sound fuel pump can fail these tests if the electrical circuit is faulty. Voltage drop tests on the fuel pump will insure that the pump has a proper power feed and a good ground connection. A DSO with a low amps probe can provide a current waveform that shows the integrity of the brushes and commutator in the motor as well as the current draw of the pump motor. Testing of known good systems to determine normal values will allow a technician to quickly pinpoint problems with a current probe and scope.

Task D.12   Determine root cause of failures.

Task D.13   Determine root cause of multiple component failures.

Task D.14   Determine root cause of repeated component failures.

Tasks D.12, D.13, and D.14 are related to each other and to similar tasks listed for other subsystems. A well-documented problem is with repeated fuel pump failures. After several replacements of good quality fuel pumps it is determined that debris in the fuel tank is getting into the fuel pumps and damaging the pump. After removing and properly cleaning the tank, the problem disappears. A sticking fuel-pressure regulator can increase fuel pressure high enough to damage the fuel pump.

Failure to test fuel pressure after replacing the pump will prevent the technician from finding the root cause of the problem, a bad pressure regulator. Problems such as these illustrate the need to test the system after repairs are made instead of only testing during the diagnostic process.

E. Emission-Control Systems Diagnosis (10 Questions)

Task E.1   Inspect and test for missing, modified, inoperative, or tampered components.

Some of the emission-control systems and devices most susceptible to modification and tampering are:

- *Air-injection systems*—The air pump may be removed entirely, or only the belt may be removed. Air-injection nozzles are sometimes plugged along with removal of the pump, hoses, and belt. Small pipe plugs screwed into exhaust manifold ports and crankshaft pulleys with no belts attached are frequent signs of missing air-injection systems.

- *EGR systems*—EGR valves are sometimes removed and their ports sealed with blank flanges or plugs. Some EGR valves can be disabled by hitting the top of the valve with a heavy hammer to increase internal spring tension and keep the diaphragm from opening the valve. Vacuum lines may be plugged with small bolts and ball bearings. Vacuum solenoids may be disconnected or removed. EGR tampering not only increases NO_x emissions, it can lead to serious engine detonation and damage because EGR is the principal means used on late-model engines to control pinging.

Many emission inspection programs require functional (operational) tests of air injection and EGR systems. These functional tests often reveal tampering that simple inspection cannot detect.

Task E.2   Locate relevant service information.

Service information for emission-control systems originates with the vehicle manufacturers, as does all other service information previously discussed. The kinds and sources of information are related to those for fuel, ignition, and general control system diagnosis and repair.

Two of the most important kinds of service information for emission-control testing and service are a component locator manual and an emission-application manual. An emission-application manual tells you what kinds of emission-control systems and devices a vehicle was originally equipped with. A component locator manual tells you where they are installed and what they look like.

Additionally, vacuum hose diagrams are essential for emission service. That is why the underhood decal contains the basic vacuum hose routing diagram for the emission devices. Remember, however, that the diagram on the decal is not the complete vacuum system for the vehicle. That is why an additional vacuum diagram manual also is necessary.

Task E.3   Research system operation using technical information to determine diagnostic procedure.

Technical information related to emission-control systems will be application tables listing systems in use, theory of operation, system functional tests, and component replacement procedures. With this information and a clear description of the problem, the technician can proceed with a systematic approach to diagnosis. While testing emission-control sub-systems is a part of any repair where the complaint was a failed emission test, it should be understood that malfunctioning emission-control systems can cause a host of drivability problems as well. Late-model emission controls are a part of the power plant design and not just hung-on afterthoughts. A problem with the EGR system can cause not only increased NO_x emissions, but also engine misfire or hesitation, rough idle and stalling and engine damage from detonation. It should be clear that proper operation of emission-control systems is essential.

Task E.4

Use the appropriate diagnostic procedures based on available vehicle data and service information; determine if available information is adequate to proceed with effective diagnosis.

The analytical methods and thought processes for effectively diagnosing emission-control systems is similar to other systems already discussed. An effective diagnosis is very difficult without proper test procedures and a clear understanding of system operation. Manufacturer specific test procedures are necessary to follow due to the large design differences among the many carmakers. Attempting to diagnose EGR or EVAP systems without vehicle specific test procedures is difficult because these systems not only vary between carmakers, but also because there may be differences among the same manufacturer's carlines.

Task E.5

Establish relative importance of displayed scan tool data.

All of the data parameters described for ignition, fuel, air intake, and general computer system operation can affect vehicle emissions and be symptoms of emission problems.

The following specific data parameters, however, are common ones for specific emission subsystems:

- *Air-injection valve operation*—Most control systems that transmit serial data include switch parameters (on-off or yes-no) that indicate the operation of air switching and air diverter valves. These data items will principally tell you if air is routed to the exhaust manifold for engine warm-up, directed downstream to the catalyst, or diverted to the atmosphere.

- *EGR system operation*—Many different data parameters indicate EGR valve operation. Some systems provide switch signals (on-off or yes-no) for EGR solenoids. Some systems transmit duty-cycle readings for pulse-width-modulated EGR valves. Others provide position sensor readings as a percentage of valve opening.

- *EVAP system control data*—Many different data parameters indicate operation of EVAP system components. Most often, switch parameters are used to indicate operation of purge and vent solenoids.

Vehicle manufacturers' system descriptions and test procedures are particularly important for determining normal and abnormal indications for many of these emission data parameters.

Task E.6

Differentiate between emission-control systems mechanical and electrical/electronic problems.

Because so many emission-control mechanical systems are electronically controlled, it is very important to differentiate between electrical or mechanical problems. A trouble code may be set for an out of range EGR valve position sensor leading the technician to believe the vehicle has an electrical problem when something as simple as a piece of carbon may be holding the valve open and causing a higher than normal sensor signal. Computer scan data may indicate that the air-injection system is delivering air to the catalytic converter, but a melted air switching valve can cause the air to be sent to the exhaust manifold instead. A failed exhaust check valve that allowed exhaust gas to flow back into the switching valve may be the root cause of the problem.

Note: Tasks seven through eleven refer to the following emission-control subsystems: positive crankcase ventilation, ignition timing control, idle and deceleration speed control, exhaust gas recirculation, catalytic converter system, secondary air-injection system, intake air temperature control, early fuel evaporation control, and evaporative emission control (including ORVR).

Task E.7

Determine need to diagnose emission-control subsystems.

Customer complaints of emission test failures, performance complaints such as pinging or spark knock, or visual inspections can all lead the technician to the need to diagnose emission-control subsystems. A technician performing routine maintenance that finds a puddle of oil in the air cleaner housing needs to test the positive crankcase ventilation system. A carbureted vehicle with a cold engine hesitation or slow warm-up complaint needs to have the early fuel evaporation system checked for proper operation. These are just two examples of how a technician may be directed toward emission-control subsystem diagnosis.

Task E.8   **Perform functional tests on emission-control subsystems; determine needed repairs.**

Functional testing of emission-control subsystems requires the technician to test all aspects of a particular system. Functional testing of the EGR system includes manually lifting the EGR valve diaphragm to see if the engine falters or stalls (indicating that EGR gases flowed into the intake) and the EGR passages are clear, checking for the presence of vacuum at the valve on systems operated by vacuum, testing the vacuum controls for proper operation and the hoses for leaks or obstructions, and monitoring any feedback sensors for correct signals. A secondary air-injection system must be tested to insure that air is delivered to the exhaust manifolds in open-loop operation, that the air switches to the catalytic converter in closed loop and that air is diverted to atmosphere during deceleration conditions. A test with an infrared gas analyzer will confirm the presence of oxygen in the exhaust stream to verify no restrictions are present and that the pump output is sufficient. A catalytic converter should be tested for a broken monolith by tapping the converter shell with a rubber mallet, rattling indicates a broken monolith and converter replacement. An exhaust back pressure test should be performed to test for restrictions. A delta temperature test with a temperature probe can be used to see if the converter will light off. Cranking CO_2 tests, snap throttle oxygen storage tests, or propane conversion tests can be done to test converter efficiency.

Task E.9   **Determine effect on exhaust emissions caused by a failure of an emission control component or subsystem.**

This task requires that the technician have a good understanding of the theory of operation for each emission subsystem and its job in controlling tailpipe emissions and also what effect the system can have on engine performance when there is a malfunction. For instance, the EGR system primarily controls NO_x emissions by lowering combustion chamber temperatures below 2500°F, but if the EGR valve sticks open, it can cause a density misfire in the engine that will raise hydrocarbons. A failure in the secondary air-injection system may cause air to be delivered upstream ahead of the oxygen sensor continuously. This problem would fool the computer into thinking the fuel control is too lean, and the computer will respond by sending a constant rich command. The increased fuel delivery will cause a rise in CO emissions and a possible emission test failure. A catalytic converter failure can cause an increase in all the harmful tailpipe emissions, due to the fact that modern three-way catalysts control HC, CO, and NO_x emissions. All other emission subsystems must be checked and in good working order for the catalytic converter to work efficiently. A catalytic converter cannot reduce hydrocarbons to an acceptable level if misfire is present in the engine, and the EGR system must reduce formation of NO_x in the combustion chamber so that the converter can bring the level low enough to pass enhanced emission inspections.

Task E.10   **Use exhaust gas analyzer readings to diagnose the failure of an emission-control component or subsystem.**

Using a four- or five-gas analyzer, the technician will perform tests to baseline a vehicle with an emission problem and verify any repairs that are completed. A four-gas analyzer will be useful for problems causing an increase in HC or CO emissions but a portable five-gas analyzer is needed for NO_x emissions problems. Due to the fact that NO_x emissions are primarily produced under road load conditions, a stationary engine analyzer with NO_x measurement capabilities is not very effective for diagnosing NO_x emissions problems. Using tables F1 and F2 located in the next section of this overview, a technician can develop a systematic approach to troubleshooting emission component or subsystem problems based on the readings gathered with a gas analyzer.

Task E.11 **Diagnose hot or cold no-starting, hard-starting, stalling, engine misfire, poor drivability, incorrect idle speed, poor idle, flooding, hesitation, surging, spark knock, power loss, poor mileage, dieseling, illuminated MIL, and emission problems caused by a failure of emission-control components or subsystems.**

Refer to Tables F1 and F2 in the next section of this Overview. These tables summarize how HC, CO, and NO_x emissions relate to various engine operating conditions and common problems. They also summarize how CO_2 and O_2 present in the exhaust will change under different circumstances.

When any of the gas measurements is out of the normal range, understanding the relationships of these exhaust gases will direct you to general area tests of the emission-control components and subsystems listed above.

Task E.12 **Determine root cause of failures.**

Task E.13 **Determine root cause of multiple component failures.**

Task E.14 **Determine root cause of repeated component failures.**

Tasks E.12, E.13, and E.14 are related to each other and to similar tasks listed for other subsystems. Replacing a failed catalytic converter without testing for other engine or emission system problems can lead to an expensive comeback. A spark plug wire arcing to the cylinder head can raise HC emissions and damage a catalytic converter. Newer vehicles equipped with OBD II certified emissions systems monitor for engine misfire to prevent this type of problem. On pre-OBD II systems it is the job of the technician to try and determine why a component failed. The arcing spark plug wire mentioned above could be the root cause of the catalytic converter failure as well as the source for repeated failures of the catalytic converter.

Task E.15 **Verify effectiveness of repairs**

Once any repair is completed the effectiveness of the repair must be verified. This operation may involve a test drive or the use of diagnostic equipment. Whichever method you choose, always make sure to verify that the repair successfully fixed the original complaint.

F. I/M Failure Diagnosis (8 Questions)

Task F.1 **Inspect and test for missing, modified, inoperative, or tampered components.**

Changing, bypassing, or tampering with electronic engine and emissions control components violates federal law. Part of troubleshooting an I/M failure is to inspect it for missing, damaged, or tampered components, just as you would inspect any other vehicle system. Refer to vehicle specific information on emission system components prior to inspecting the system.

A good visual inspection should be your first step when diagnosing I/M failures.

Regarding air-injection systems: Most newer vehicles no longer have an air pump to check. Some vehicles with V-8 engines are equipped with an electric air pump controlled by the PCM. This air pump operates only during cold startup and WOT. Refer to Emission-Control Systems Diagnosis.

Task F.2 **Locate relevant service information.**

Service information for emission-control systems originates with the vehicle manufacturers, as does all other service information previously discussed. The kinds and sources of information are related to those for fuel, ignition, and general control system diagnosis and repair.

Two of the most important kinds of service information for emission-control testing and service are a component locator manual and an emission-application manual. An emission-application manual tells you what kinds of emission-control systems and devices a vehicle was originally equipped with. A component locator manual tells you where they are installed and what they look like.

Additionally, vacuum hose diagrams are essential for emission service. That is why the underhood decal contains the basic vacuum hose routing diagram for the emission devices. Remember, however, that the diagram on the decal is not the complete vacuum system for the vehicle. That is why an additional vacuum diagram manual also is necessary.

Task F.3   **Evaluate emission readings obtained during an I/M test to assist in emission failure diagnosis and repair.**

Determining which operational mode the vehicle was in during the failure is a very important step in your diagnosis. Knowing the operational mode narrows the possible causes of the failure. During certain operational modes, the processor does not look at some input devices and at other times the processor is in a fixed or modified fuel and timing delivery mode. The various operational modes will change from vehicle to vehicle and will require vehicle specific information.

Using the readings obtained from an I/M test will provide the technician with the emissions failure, how much it failed by and which mode it failed in. Using this information will help the technician decide the proper diagnostic and repair procedures.

For example: If a vehicle fails the I/M test for high NO_x during acceleration only.

Knowing how the NO_x is reduced and by which component, would lead you to the EGR valve operation. If the EGR is controlled by the PCM, knowing the operational mode during failure may help you pinpoint the problem. If the vehicle failed NO_x at all times, the number of systems and components to be checked would increase.

Task F.4   **Evaluate HC, CO, NO_x, CO_2, O_2 gas readings; determine the failure relationships.**

Tables F1 and F2 summarize how HC, CO, NO_x, CO_2, and O_2 emissions relate to various engine operating conditions and common problems. Refer to these tables to review the failure relationships. The Vehicle Inspection Report (VIR) provided to the vehicle owner if the vehicle fails an enhanced IM240 emission test is extremely important because it displays the vehicle's emissions output on a graph so the technician can see how the operating conditions affected the emissions. The vehicle drive trace is superimposed on each exhaust gas trace so the technician can compare emissions produced to how the vehicle was being driven, such as idling, accelerating, decelerating, or steady cruise conditions. Elevated emissions outputs under certain driving conditions can point the technician in the proper direction for diagnosis of many problems. For instance, a sticking EGR valve can cause HC spikes during decel conditions; a saturated charcoal canister may cause elevated CO emissions during steady state cruise when purge is commanded on; and a weak catalytic converter may produce a drive trace showing elevated HC, CO, and NO_x during the high speed portion of the drive trace when exhaust flow is highest.

Task F.5   **Use test instruments to observe, recognize, and interpret electrical/electronic signals.**

The principal test instruments used for I/M emission testing are the four- or five-gas exhaust analyzer and, for loaded-mode testing, a dynamometer. Additionally, the same arsenal of test equipment used for any part of engine performance troubleshooting and service should be available. Electrical and electronic test equipment includes ignition oscilloscopes, laboratory oscilloscopes (lab scopes), digital volt-ohm-ammeters (DVOM), and scan tools.

Voltage-drop testing discussed in section C of this overview is as important for I/M testing as for any electrical troubleshooting. Ignition scope tests can reveal ignition problems; but just as importantly, abnormalities in ignition voltage waveforms can indicate fuel system and mechanical problems.

Lab scope testing has an advantage over basic voltmeter testing because an oscilloscope paints a picture of voltage changes over a period of time. The two-dimensional waveform, or trace, can reveal irregularities (glitches) that a voltmeter would not indicate.

Waveform analysis of the oxygen sensor with a labscope is one of the most important tests that should be done when diagnosing emission test failures. An oxygen sensor waveform can provide information such as whether or not misfire is present during driving conditions and whether or not the computer is in proper fuel control. A scan data reading of closed-loop does not mean the system is in fuel control, only that the conditions necessary for closed-loop operation have been met. A proper oxygen sensor waveform provides confirmation of normal fuel control. Normal oxygen sensor switching frequency is critical to proper catalytic converter operation. A properly operating fuel control system on a fuel-injected engine will show an oxygen-sensor switching rate between 0.5 to 5 Hz at 2500 rpm. Oxygen sensor switching frequency will decrease at low rpm or idle conditions and should always be checked at 2000 to 2500 rpm when comparing to specifications.

When testing an oxygen sensor for proper activity with a labscope, it is easiest to calculate switching frequency by setting the scope time-base to display ten seconds of time across the scope screen. Using this setting, if the oxygen sensor switches 10 times during one screen sweep the frequency is 1 Hz, if 18 switches are shown the frequency is 1.8 Hz. Slow or lazy oxygen sensors throw off fuel control and reduce converter efficiency. This problem may not be apparent if the oxygen sensor is tested with a voltmeter but is clearly visible when tested with a labscope.

Task F.6   Analyze HC, CO, NO$_x$, CO$_2$, and O$_2$ readings; determine diagnostic test sequence.

Tables F1 and F2 summarize how HC, CO, NO$_x$, CO$_2$, and O$_2$ emissions relate to various engine operating conditions and common problems. Refer to these tables to review the failure relationships.

When you understand a particular combination of exhaust analyzer readings and emissions that exceed test limits, you can choose the most appropriate test and repair sequences based on the most probable causes listed in Tables F1 and F2. These procedures most often will be the ones you have reviewed in previous sections of this overview.

Task F.7   Diagnose the cause of no-load I/M test HC emission failures.

Hydrocarbons are unburned fuel molecules. No-load I/M test HC emission failures indicate a problem with combustion efficiency. Excessive HC present in the exhaust can be caused by anything that affects the combustion process. Items to be considered for the cause of high HC include cylinder compression and airflow, ignition operation and timing, air/fuel mixtures, and possible engine modifications or tampering.

Task F.8   Diagnose the cause of no-load I/M test CO emission failures.

Carbon monoxide present in the exhaust stream indicates there was excessive fuel present for the amount of air in the cylinder during combustion. This is primarily a rich mixture indicator. Review Table F1 headings "Very Rich—Below 10:1 at all speeds" and "Rich—10:1 to 12:1 at low speed only" for a listing of common causes for excessive CO emissions. Most no-load I/M test CO emission failures can be traced to one of the causes listed there.

Task F.9   Diagnose the cause loaded-mode I/M test HC emission failures.

Causes for HC failures during loaded-mode I/M testing are the same as for no-load testing but the engine operating conditions are different and thus diagnostic strategies must be changed. Problems that may not show up under static engine testing can become an issue when the engine is operated under load. An arcing spark plug wire or lean injector may not cause a misfire until the engine is loaded; these types of problems need to be identified. Oxygen sensor waveform analysis is one good method for determining if this type of problem exists. Portable gas analyzers also become much more useful in tracking down emission problems that require testing in a loaded condition.

Task F.10   Diagnose the cause loaded-mode I/M test CO emission failures.

Strategies for diagnosing the cause of loaded-mode I/M test CO emission failures are similar to those discussed in Task F.9. For instance, many vehicles may not command charcoal canister purge until the computer sees a vehicle speed input. Exhaust gas testing in the service bay may show normal CO readings leading the technician to believe the engine is operating properly. A portable gas analyzer can show high CO output from the canister purging because measurements are taken under normal driving conditions. The technician needs to understand how the various systems operate and how different driving conditions can cause problems to occur.

Task F.11   Diagnose the cause loaded-mode I/M test NO$_x$ emission failures

Loaded-mode I/M test NO$_x$ emission failures are one of the most challenging emission test failures for technicians to diagnose. Most previous state emission test programs did not test NO$_x$ emissions, so many technicians have not had experience in troubleshooting causes for increased NO$_x$ in the exhaust. Another problem is the fact that many shops do not possess five-gas analyzers, so they have no means of measuring NO$_x$ and determining whether or not any repairs performed were effective. Many engine systems can contribute to excessive NO$_x$ emissions.

While EGR is the single most important system in controlling NO$_x$, problems in the cooling system that raise engine-operating temperature, over-advanced ignition timing, or a failed catalytic converter will all contribute to increased NO$_x$ production. Even correcting a different emission problem can create an increase in NO$_x$ levels. Carbon deposits formed in the combustion chamber from running an engine with an excessively rich air/fuel mixture can increase compression pressure. Repairing the rich running condition will lean the air/fuel mixture, but the remaining high compression from the carbon deposits can cause NO$_x$ levels to go up. Performing a de-carbon treatment to the engine prior to an emission re-test is recommended in this case. This condition has caused many emission re-test failures. Base lining a vehicle with a five-gas analyzer at the beginning of diagnostics and repeating the test after repairs to verify the effectiveness of any repairs is necessary to prevent the vehicle from failing a re-test.

Task F.12   Evaluate the MIL operation for onboard diagnostic I/M testing.

The ODBII system will be checked on 1996 and later year vehicles during an inspection and maintenance (I/M) test. The system is checked for emission-related diagnostic trouble codes. The vehicle fails the enhanced I/M test if any emission-related fault codes are found and the MIL is commanded on.

Task F.13   Evaluate monitor readiness status for onboard diagnostic I/M testing.

Some vehicles require a very specific combination of temperature, speed, and load changes to be included in a drive cycle in order to trigger a particular monitor. The drive cycle changes from one monitor to another and one vehicle to another. This is why good service information on the particular vehicle is essential. Monitor readiness can be observed with the use of a scan tool.

Task F.14   Diagnose communication failures with the vehicle during onboard diagnostic I/M testing.

During an I/M test the vehicle is checked for any emission-related diagnostic trouble codes using a scan tool. The data link connector must be functional in order to proceed with the I/M test. A failure to communicate with the PCM during an I/M test will fail a vehicle even if it passed the tail pipe emissions test.

Task F.15   Perform functional I/M tests (including fuel cap tests).

Inspection and Maintenance (I/M) programs require vehicle emission systems inspections. Emission system inspections ensure that the emission system operates properly. If a vehicle is emitting excessive emissions, the owner must have the emission system repaired to meet emission standards. An emission gas analyzer is used to measure tail pipe emissions. An enhanced I/M test is done on a dynamometer. The five major inspections performed during the enhanced I/M test include a visual inspection, tail pipe emissions on a dynamometer, evaporative purge flow, evaporative leak test, and OBD II systems check for 1996 and newer.

Task F.16   Verify effectiveness of repairs.

Upon completion of the repairs, the vehicle should be re-tested and the results compared to the first test when the vehicle failed. You should see improvement in your test results, the bigger the improvement the more effective the repair.

Some portable gas analyzers allow before and after emission snapshots to be taken, similar to snapshots that are done with a scan tool. The analyzer will compare two snapshots and display the actual change in percentage of the emissions measured. Using this feature, the technician can determine if the emissions improved or got worse and by how much. This feature may also be available by connecting the gas analyzer with a PC and dedicated software created by the equipment manufacturer. In either case, a direct means of measuring repair effectiveness is possible.

Table F1
EXHAUST EMISSIONS

AIR/FUEL RATIO	Engine Speed	HC	CO	CO_2	O_2
Very Rich—Below 10:1 at all speeds	Idle	250 ppm	3%	7 to 9%	0.2%
	Off Idle	275 ppm	3%	7 to 9%	0.2%
	Cruise	300 ppm	3%	7 to 9%	0.2%

Other Symptoms Black smoke or sulfur odor, poor fuel economy, surge or hesitation, stalling, rough or "lumpy" idle, engine not warming to operating temperature, continuous open-loop operation

Possible Causes

- High MAP sensor voltage (vacuum leak or electrical fault)
- Leaking fuel injectors
- High fuel pressure
- High float level or leaking power valve in a carburetor
- Thermostat stuck open or engine otherwise continuously operating at very low temperature

EXHAUST EMISSIONS

AIR/FUEL RATIO	Engine Speed	HC	CO	CO_2	O_2
Rich—10:1 to 12:1 at low speed only	Idle	150 ppm	1.5%	7 to 9%	0.5%
	Off Idle	150 ppm	1.5%	7 to 9%	0.5%
	Cruise	100 ppm	1.0%	11 to 13%	1.0%

Other Symptoms Poor fuel economy, surge or hesitation, black smoke and soot-fouled spark plugs, rough idle, vapor canister saturated with fuel or purge valve bad

Possible Causes

- High MAP sensor voltage (vacuum leak or electrical fault)
- Leaking fuel injectors
- Engine oil diluted with gasoline
- Excessive crankcase blowby
- High fuel pressure
- High float level or leaking power valve in a carburetor
- Carburetor idle speed too low, mixture too rich, or choke stuck closed
- Thermostat stuck open or engine otherwise continuously operating at very low temperature

EXHAUST EMISSIONS

AIR/FUEL RATIO	Engine Speed	HC	CO	CO_2	O_2
Very Lean—Above 16:1 at all speeds	Idle	200 ppm	0.5%	7 to 9%	4 to 5%
	Off Idle	205 ppm	0.5%	7 to 9%	4 to 5%
	Cruise	250 ppm	1.0%	7 to 9%	4 to 5%

Other Symptoms Rough idle, high-speed misfire, overheating, surging, hesitation, detonation at cruising speeds

Table 1, cont'd.

Possible Causes

- Intermittent ignition problems causing misfire
- Restricted fuel injectors
- Low fuel pressure
- Vacuum leak
- Low carburetor float level or lean carburetor mixture
- Poor cylinder sealing (low compression)
- Improper ignition timing
- Thermostat stuck closed or engine otherwise operating at very high temperature

EXHAUST EMISSIONS

AIR/FUEL RATIO	Engine Speed	HC	CO	CO_2	O_2
Lean—Above 16:1 at high speed	Idle	100 ppm	2.5%	7 to 9%	2 to 3%
	Off Idle	80 ppm	1.0%	7 to 9%	2 to 3%
	Cruise	50 ppm	0.8%	7 to 9%	2 to 3%

Other Symptoms　Rough idle, misfire, surging, hesitation

Possible Causes

- Intermittent ignition problems causing misfire
- Restricted fuel injectors
- Low fuel pressure
- Vacuum leak
- Low carburetor float level or lean carburetor mixture
- Carburetor heated air intake stuck in cold-air position

EXHAUST EMISSIONS

AIR/FUEL RATIO	Engine Speed	HC	CO	CO_2	O_2
Normal—13:1 to 15:1 but engine not fully warmed up	Idle	100 ppm	0.3%	10 to 12%	2.5%
	Off Idle	80 ppm	0.3%	10 to 12%	2.5%
	Cruise	50 ppm	0.3%	10 to 12%	2.5%

Other Symptoms　Cold-engine emission test failure, Catalytic converter not warmed up

Table F1: *This table illustrates the relationships of exhaust gas measurements on a four-gas analyzer to different air/fuel ratios. Studying this table will help you understand how emissions change in different ways as a result of different air/fuel ratios. You also can use this table to guide you toward probable causes of certain abnormal combinations of exhaust gas measurements.*

Table F2
EXHAUST EMISSIONS

ENGINE PROBLEM	HC	CO	CO_2	O_2	NO_x
Rich Mixture	Moderate Increase	Large Increase	Some Decrease	Some Decrease	Moderate Decrease
Lean Mixture	Moderate Increase	Large Decrease	Some Decrease	Some Increase	Moderate Increase
Very Lean Mixture	Large Increase	Large Decrease	Some Decrease	Large Increase	Large Increase
Ignition Misfire	Large Increase	Some Decrease	Some Decrease	Moderate Increase	Moderate Decrease
Advanced Timing	Some Increase	No Change or Slight Decrease	No Change	No Change	Large Increase
Retarded Timing	Some Decrease	No Change or Slight Increase	No Change	No Change	Large Decrease
Very Retarded Timing	Some Increase	No Change	Moderate Decrease	No Change	Some Increase
Low Compression	Moderate Increase	Some Decrease	Some Decrease	Some Increase	Moderate Decrease
Exhaust Leak	Some Decrease	Some Decrease	Some Decrease	Some Increase	No Change
Worn (Flat) Cam Lobes	No Change or Some Decrease	Some Decrease	Some Decrease	No Change or Some Decrease	No Change or Some Decrease
General Engine Wear	Some Increase	Some Increase	Some Decrease	Some Decrease	No Change or Slight Decrease
Air Injection Failure	Some Increase	Large Increase	Moderate Decrease	Moderate Decrease	No Change
EGR Leaking	Some Increase	No Change	No Change or Some Decrease	No Change	No Change or Some Decrease
NORMAL EMISSION CONTROL CONDITION	HC	CO	CO_2	O_2	NO_x
EGR Operating Normally	No Change	No Change	Some Decrease	No Change	Large Decrease
Air Injection Operating Normally	Large Decrease	Large Decrease	Moderate Decrease	Large Increase	No Change

Table F2: This table illustrates the relationships of exhaust gas measurements on a five-gas analyzer to different engine conditions. Studying this table will help you understand how emissions change in different way as a result of different engine problems. You also can use this table to guide you toward probable causes of certain abnormal combinations of exhaust gas measurements.

Composite Vehicle

1. You will notice as you read through the Task List that the job skills identified concentrate on the ability to diagnose, rather than repair. The panel of experts who developed the L1 test have identified three important general characteristics of drivability diagnosis.
2. Data is obtained from the vehicle using a variety of test instruments and is compared to known values obtained from the service manuals.
3. A good technician can draw valid conclusions from the relationship between published data and what he understands of the vehicles fuel, ignition, and emission-control systems.

The ASE panel of experts has developed a "Composite Vehicle" engine control system, which is described in detail in the "Composite Vehicle Preparation/Registration Booklet" you'll receive when you register for the test. The information is included here so that you can begin the familiarization process now.

The composite vehicle uses a "mass airflow" fuel-injection system design used by many domestic and import manufacturers. The system uses sensor, actuators, and control strategies that you should be familiar with from your shop experience. When you answer question based on the Composite Vehicle, you will be simulating your real-world experience of using reference materials and test instruments to diagnose problems based on your knowledge of a particular engine-management system.

The following Composite Vehicle Information has been provided by the National Institute for Automotive Service Excellence (ASE). Thomson Delmar Learning would like to thank the ASE for providing this content for use in this study guide. To receive a copy of ASE's Composite Vehicle Reference booklet please contact the ASE at:

National Institute for Automotive Service Excellence
101 Blue Seal Drive S.E.
Suite 101
Leesburg, VA 20175
Telephone: 703-669-6600
www.ase.com

GENERAL DESCRIPTION

This generic four cycle, V6 engine has four overhead chain-driven camshafts, 24 valves, distributorless ignition, and a mass airflow-type closed-loop sequential multiport fuel injection system. The Engine Control Module (ECM) receives input from sensors, calculates ignition and fuel requirements, and controls engine actuators to provide the desired driveability, fuel economy, and emissions control. The ECM also controls the vehicle's charging system. The powertrain control system has OBD II-compatible sensors and diagnostic capabilities. The ECM receives power from the battery and ignition switch and provides a regulated 5 volt supply for most of the engine sensors. The engine is equipped with a single exhaust system and a three-way catalytic converter, without any secondary air injection. Engine control features include variable valve timing, electronic throttle actuator control (TAC), a data communications bus, a vehicle anti-theft immobilizer system, and on-board refueling vapor recovery (ORVR) EVAP components. The control system software and OBD II diagnostic procedures stored in the ECM can be updated using factory supplied calibration files and PC-based interface software, along with a reprogramming device or scan tool that connects the PC to the vehicle's data link connector (DLC).

AUTOMATIC TRANSMISSION

- Four-speed automatic overdrive transaxle, with shifting controlled by a transmission control module (TCM). The TCM communicates with the ECM and other modules through the data bus.
- The TCM provides its own regulated 5 volt supply, performs all OBD II transaxle diagnostic routines, and stores transaxle diagnostic trouble codes (DTCs). The control system software and OBD II diagnostic procedures stored in the TCM can be updated in the same way as the ECM.
- Failures that result in a pending or confirmed DTC related to any of the following components will cause the TCM to default to fail-safe mode: range switch, shift solenoids, turbine shaft speed sensor, and the vehicle speed sensor. The TCM will also default to fail-safe mode if it is unable to communicate with the ECM.
- When in fail-safe mode, maximum line pressure will be commanded, the transmission will default to 2nd gear, and the torque converter clutch will be disabled.

FUEL SYSTEM

- Sequential Multiport Fuel Injection (SFI)
- Returnless Fuel Supply with electric fuel pump mounted inside the fuel tank
- Fuel pressure is regulated to a constant 50 psi (345 kPa) by a mechanical regulator in the fuel tank. Minimum acceptable fuel pressure is 45 psi (310 kPa). The fuel system should maintain a minimum of 45 psi (310 kPa) for two minutes after the engine is turned off.

IGNITION SYSTEM

- Distributorless Ignition (EI) with six ignition coils (coil-on-plug)
- Firing Order: 1-2-3-4-5-6
- Cylinders 1, 3, and 5 are on Bank 1, Cylinders 2, 4, and 6 are on Bank 2
- Ignition timing is not adjustable
- Timing is determined by the ECM using the crankshaft position (CKP) sensor signal
- The ignition coil drivers are integrated into the ECM

VARIABLE VALVE TIMING

- Intake camshaft timing is continuously variable using a hydraulic actuator attached to the end of each intake camshaft. Engine oil flow to each hydraulic actuator is controlled by a camshaft position actuator control solenoid. Exhaust camshaft timing is fixed.

- A single timing chain drives both exhaust camshafts and both intake camshaft hydraulic actuators. While valve overlap is variable, valve lift and duration are fixed.
- Cam timing is determined by the ECM using the crankshaft position (CKP) sensor and camshaft position sensor (CMP 1 and CMP 2) signals. At idle, the intake camshafts are fully retarded and valve overlap is zero degrees. At higher speeds and loads, the intake camshafts can be advanced up to 40 crankshaft degrees.
- Each intake camshaft has a separate camshaft position sensor, hydraulic actuator, and control solenoid. If little or no oil pressure is received by a hydraulic actuator (typically at engine startup, at idle speed, or during a fault condition), it is designed to mechanically default to the fully retarded position (zero valve overlap), and is held in that position by a spring-loaded locking pin.

ELECTRONIC THROTTLE CONTROL

- The vehicle does not have a mechanical throttle cable, a cruise control throttle actuator, or an idle air control (IAC) valve. Throttle opening at all engine speeds and loads is controlled directly by a throttle actuator control (TAC) motor mounted on the throttle body housing.
- Dual accelerator pedal position (APP) sensors provide input from the vehicle operator, while the actual throttle angle is determined using dual throttle position (TP) sensors.
- If one APP sensor or one TP sensor fails, the ECM will turn on the malfunction indicator lamp (MIL) and limit the maximum throttle opening to 35%. If any two (or more) of the four sensors fail, the ECM will turn on the MIL and disable the electronic throttle control.
- In case of failure of the electronic throttle control system, the system will default to limp-in operation. In limp-in mode, the spring-loaded throttle plate will return to a default position of 15% throttle opening, and the TAC value on the scan tool will indicate 15%. This default position will provide a fast idle speed of 1400 to 1500 rpm, with no load and all accessories off.
- Normal no-load idle range is 850 to 900 rpm at 5% to 10% throttle opening.
- No idle relearn procedure is required after component replacement or a dead battery.

DATA COMMUNICATIONS BUS

- The serial data bus is a high-speed, non-fault tolerant, two wire twisted pair communications network. It allows peer-to-peer communications between various electronic modules, including the engine control module (ECM), transmission control module (TCM), instrument cluster (including the MIL), immobilizer control module, and a scan tool connected to the data link connector (DLC).
- The Data-High circuit switches between 2.5 (rest state) and 3.5 volts (active state), and the Data-Low circuit switches between 2.5 (rest state) and 1.5 volts (active state). The data bus has two 120 ohm terminating resistors: one inside the instrument cluster, and another one inside the ECM.
- Any of the following conditions will cause the data communications bus to fail and result in the storage of network DTCs: either data line shorted to power, to ground, or to the other data line.
- The data bus will remain operational when one of the two modules containing a terminating resistor is not connected to the network. The data bus will fail when both terminating resistors are not connected to the network.
- Data communication failures do not prevent the ECM from providing ignition and fuel control.

IMMOBILIZER ANTI-THEFT SYSTEM

- When the ignition switch is turned on, the immobilizer control module sends a challenge signal through the antenna around the ignition switch to the transponder chip in the ignition key. The transponder key responds with an encrypted key code. The immobilizer control module then decodes the key code and compares it to the list of registered keys.
- When the engine is started, the ECM sends a request to the immobilizer control module over the data bus to verify the key validity. If the key is valid, the immobilizer control module responds with

a "valid key" message to the ECM to continue normal engine operation.

- If an attempt is made to start the vehicle with an invalid ignition key, the immobilizer control module sends a message over the data bus to the instrument cluster to flash the anti-theft indicator lamp. If the ECM does not receive a "valid key" message from the immobilizer control module within 2 seconds of engine startup, the ECM will disable the fuel injectors to kill the engine. Cycling the key off and cranking the engine again will result in engine restart and stall.
- The immobilizer control module and ECM each have their own unique internal ID numbers used to encrypt their messages, and are programmed at the factory to recognize each other. If either module is replaced, the scan tool must be used to program the replacement module, using the VIN, the date, and a factory-assigned PIN number.
- Up to eight keys can be registered in the immobilizer control module. Each key has its own unique internal key code. If only one valid key is available, or if all keys have been lost, the scan tool can be used to delete lost keys and register new keys. This procedure also requires the VIN, the date, and a factory-assigned PIN number.
- Neither the ECM, TCM, nor the immobilizer control module prevent operation of the starter motor for anti-theft purposes.

ON-BOARD REFUELING VAPOR RECOVERY (ORVR) EVAP SYSTEM

- The on-board refueling vapor recovery EVAP system causes fuel tank vapors to be directed to the EVAP charcoal canister during refueling, so that HC vapors do not escape into the atmosphere.
- The following components have been added to the traditional EVAP system for ORVR capability: a one inch I.D. fill pipe, a one-way check valve at the bottom of the fill pipe, a fuel vapor control valve inside the fuel tank, and a ½ inch I.D. vent hose from the vapor control valve to the canister.
- The fuel vapor control valve has a float that rises to seal the vent hose when the fuel tank is full. It also prevents liquid fuel from reaching the canister and blocks fuel from leaking in the event of a vehicle roll-over.

SENSORS

- ### MASS AIRFLOW (MAF) SENSOR
  Senses airflow into the intake manifold. The sensor reading varies from 0.2 volts (0 gm/sec) at key-on, engine-off, to 4.8 volts (175 gm/sec) at maximum airflow. At sea level, no-load idle (850 rpm), the sensor reading is 0.7 volts (2.0 gm/sec). Located on the air cleaner housing.

- ### MANIFOLD ABSOLUTE PRESSURE (MAP) SENSOR
  Senses intake manifold absolute pressure. The MAP sensor signal is used by the ECM for OBD II diagnostics only. The sensor reading varies from 4.5 volts at 0 in. Hg vacuum / 101 kPa pressure (key on, engine off, at sea level), to 0.5 volts at 24 in. Hg vacuum / 20.1 kPa pressure. At sea level, no-load idle with 18 in. Hg vacuum (40.4 kPa absolute pressure), the sensor reading is 1.5 volts. Located on the intake manifold.

- ### ENGINE COOLANT TEMPERATURE (ECT) SENSOR
  A negative temperature coefficient (NTC) thermistor that senses engine coolant temperature. The sensor values range from -40°F to 248°F (-40°C to 120°C). At 212°F (100°C), the sensor reading is 0.46 volts. Located in the engine block water jacket.

- ### INTAKE AIR TEMPERATURE (IAT) SENSOR
  A negative temperature coefficient (NTC) thermistor that senses air temperature. The sensor values range from -40°F to 248°F (-40°C to 120°C). At 86°F (30°C), the sensor reading is 2.6 volts. Located in the air cleaner housing.

• CRANKSHAFT POSITION (CKP) SENSOR

A magnetic-type sensor that generates 35 pulses for each crankshaft revolution. It is located on the front engine cover, with a 35-tooth reluctor wheel mounted on the crankshaft just behind the balancer pulley. Each tooth is ten crankshaft degrees apart, with one space for a "missing tooth" located at 60 degrees before top dead center of cylinder number 1.

• CAMSHAFT POSITION SENSORS (CMP 1 AND CMP 2)

A pair of three-wire solid state (Hall-effect or optical-type) sensors that generate a signal once per intake camshaft revolution. The leading edge of the bank 1 CMP signal occurs on the cylinder 1 compression stroke, and the leading edge of the bank 2 CMP signal occurs on the cylinder 4 compression stroke. When the intake camshafts are fully retarded (zero valve overlap), the signals switch at top dead center of cylinders 1 and 4. When the intake camshafts are fully advanced (maximum valve overlap), the signals switch at 40 crankshaft degrees before top dead center. These signals allow the ECM to determine fuel injector and ignition coil sequence, as well as the actual intake valve timing. Loss of one CMP signal will set a DTC, and valve timing will be held at the fully retarded position (zero valve overlap). If neither CMP signal is detected during cranking, the ECM will store a DTC and disable the fuel injectors, resulting in a no-start condition. Located at the rear of each valve cover, with an interrupter mounted on the intake camshafts to generate the signal. The following diagram shows the CKP and CMP sensor signal waveforms with the camshafts at the default (fully retarded) position.

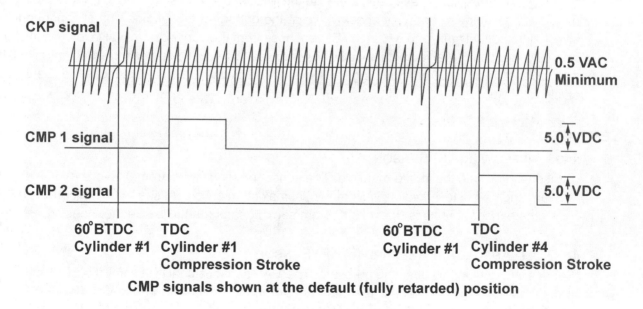

CMP signals shown at the default (fully retarded) position

• THROTTLE POSITION (TP 1 AND TP 2) SENSORS

A pair of redundant non-adjustable potentiometers that sense throttle position. The TP 1 sensor signal varies from 4.5 volts at closed throttle to 0.5 volts at maximum throttle opening (decreasing voltage with increasing throttle position). The TP 2 sensor signal varies from 0.5 volts at closed throttle to 4.5 volts at maximum throttle opening (increasing voltage with increasing throttle position). Failure of one TP sensor will set a DTC and the ECM will limit the maximum throttle opening to 35%. Failure of both TP sensors will set a DTC and cause the throttle actuator control to be disabled, and the spring-loaded throttle plate will return to the default 15% position (fast idle). Located on the throttle body.

● ACCELERATOR PEDAL POSITION (APP 1 AND APP 2) SENSORS

A pair of redundant non-adjustable potentiometers that sense accelerator pedal position. The APP 1 sensor signal varies from 0.5 volts at the released pedal position to 3.5 volts at maximum pedal depression (increasing voltage with increasing pedal position). The APP 2 sensor signal varies from 1.5 volts at the released pedal position to 4.5 volts at maximum pedal depression (increasing voltage with increasing pedal position, offset from the APP 1 sensor signal by 1.0 volt). The ECM interprets an accelerator pedal position of 80% or greater as a request for wide open throttle. Failure of one APP sensor will set a DTC and the ECM will limit the maximum throttle opening to 35%. Failure of both APP sensors will set a DTC and cause the throttle actuator control to be disabled, and the spring-loaded throttle plate will return to the default 15% position (fast idle). Located on the accelerator pedal assembly.

● EGR VALVE POSITION SENSOR

A three-wire non-adjustable potentiometer that senses the position of the EGR valve pintle. The sensor reading varies from 0.50 volts when the valve is fully closed to 4.50 volts when the valve is fully opened. Located on top of the EGR valve.

● KNOCK SENSOR

A two-wire piezoelectric sensor that generates an AC voltage spike when engine vibrations within a specified frequency range are present, indicating spark knock. The signal is used by the ECM to retard ignition timing when spark knock is detected. The sensor signal circuit normally measures 2.5 volts DC with the sensor connected.

● HEATED OXYGEN SENSORS (HO2S 1/1, HO2S 2/1, AND HO2S 1/2)

Electrically heated zirconia sensors that measure oxygen content in the exhaust stream. Sensor 1/1 is located on the Bank 1 exhaust manifold (cylinders 1, 3 and 5). Sensor 2/1 is located on the Bank 2 exhaust manifold (cylinders 2, 4, and 6). Both upstream sensor signals are used for closed loop fuel control and OBD II monitoring. Sensor 1/2 is mounted in the exhaust pipe after the catalytic converter (downstream). Its signal is used for OBD II monitoring of catalytic converter operation. The sensor outputs vary from 0.0 to 1.0 volt. When a sensor reading is less than 0.45 volts, oxygen content around the sensor is high; when a sensor reading is more than 0.45 volts, oxygen content around the sensor is low. No bias voltage is applied to the sensor signal circuits by the ECM. With the key on and engine off, the sensor readings are zero volts. Battery voltage is continuously supplied to the oxygen sensor heaters whenever the ignition switch is on. The ECM will provide the ground for both of the upstream oxygen sensor heaters as soon as the engine starts. Once the engine is started, the ECM will then provide the ground for the downstream oxygen sensor heater after two minutes of continuous engine operation.

● POWER STEERING PRESSURE (PSP) SWITCH

A switch that closes when high pressure is detected in the power steering system. The signal is used by the ECM to adjust throttle position to compensate for the added engine load from the power steering pump. Located on the P/S high pressure hose.

● BRAKE PEDAL POSITION (BPP) SWITCH

A switch that closes when the brake pedal is depressed (brakes applied). The signal is used by the TCM to release the torque converter clutch. Located on the brake pedal.

- ## A/C On/Off Request Switch

  A switch that is closed by the vehicle operator to request A/C compressor operation. Located in the climate control unit on the instrument panel.

- ## A/C Pressure Sensor

  A three-wire solid-state sensor for A/C system high-side pressure. The sensor reading varies from 0.25 volts at 25 psi to 4.50 volts at 450 psi. The signal is used by the ECM to control the A/C compressor clutch, radiator fan, and to adjust throttle position to compensate for the added engine load from the A/C compressor. The ECM will also disable compressor operation if the pressure is below 40 psi or above 420 psi. Located on the A/C high side vapor line.

- ## Fuel Level Sensor

  A potentiometer that is used to determine the fuel level. The reading varies from 0.5 volts / 0% with an empty tank to 4.5 volts / 100% with a full tank. When the fuel tank is ¼ full, the sensor reading is 1.5 volts. When the fuel tank is ¾ full, the sensor reading is 3.5 volts. Used by the ECM when testing the evaporative emission (EVAP) system. Located in the fuel tank.

- ## Fuel Tank (EVAP) Pressure Sensor

  Senses vapor pressure or vacuum in the evaporative emission (EVAP) system compared to atmospheric pressure. The sensor reading varies from 0.5 volts at 1/2 psi (14 in. H_2O) **vacuum** to 4.5 volts at 1/2 psi (14 in. H_2O) **pressure**. With no pressure or vacuum in the fuel tank (gas cap removed), the sensor output is 2.5 volts. Used by the ECM for OBD II evaporative emission system diagnostics only. Located on top of the fuel tank.

- ## Vehicle Speed Sensor (VSS)

  A magnetic-type sensor that senses rotation of the final drive and generates a signal that increases in frequency as vehicle speed increases. The TCM uses the VSS signal to control upshifts, downshifts, and the torque converter clutch. The TCM communicates the VSS signal over the data communications bus to the ECM to control high-speed fuel cutoff, and to the Instrument Cluster for speedometer operation. The signal is displayed on the scan tool in miles per hour and kilometers per hour. Located on the transaxle housing.

- ## Transmission Fluid Temperature (TFT) Sensor

  A negative temperature coefficient (NTC) thermistor that senses transmission fluid temperature. The sensor values range from -40°F to 248°F (-40°C to 120°C). At 212°F (100°C), the sensor reading is 0.46 volts. This signal is used by the TCM to delay shifting when the fluid is cold, and control torque converter clutch operation when the fluid is hot. Located in the transaxle oil pan.

- ## Transmission Turbine Shaft Speed (TSS) Sensor

  A magnetic-type sensor that senses rotation of the torque converter turbine shaft (input / mainshaft) and generates a signal that increases in frequency as transmission input speed increases. Used by the TCM to control torque converter clutch operation and sense transmission slippage. Located on the transaxle housing.

- ## Transmission Range (TR) Switch

  A six-position switch that indicates the position of the transaxle manual select lever: Park/Neutral, Reverse, Manual Low (1), Second (2), Drive (3), or Overdrive (OD). Used by the TCM to control line pressure, upshifting, and downshifting. The TR switch is NOT used to enable or disable starter motor operation. Located on the transaxle housing.

ACTUATORS

- ### FAN CONTROL (FC) RELAY

  When energized, provides battery voltage (B+) to the radiator/condenser cooling fan motor. The ECM will turn the fan on when engine coolant temperature reaches 210°F (99°C) and off when coolant temperature drops to 195°F (90°C). The ECM will also turn the fan on when the A/C high side pressure reaches 300 psi and off when the pressure drops to 250 psi. The relay coil resistance spec is 48 ± 6 ohms.

- ### FUEL PUMP (FP) RELAY

  When energized, supplies battery voltage (B+) to the fuel pump. The relay coil resistance spec is 48 ± 6 ohms.

- ### A/C CLUTCH RELAY

  When energized, provides battery voltage (B+) to the A/C compressor clutch coil. The relay coil resistance spec is 48 ± 6 ohms.

- ### THROTTLE ACTUATOR CONTROL (TAC) MOTOR

  A bidirectional pulse-width modulated DC motor that controls the position of the throttle plate. A scan tool data value of 0% indicates an ECM command to fully close throttle plate, and a value of 100% indicates an ECM command to fully open the throttle plate (wide open throttle). Any throttle control actuator motor circuit fault will set a DTC and cause the throttle actuator control to be disabled, and the spring-loaded throttle plate will return to the default 15% position (fast idle). When disabled, the TAC value on the scan tool will indicate 15%.

- ### MALFUNCTION INDICATOR LAMP (MIL)

  The MIL is part of the instrument cluster and receives commands from the ECM and TCM over the data communications bus. If the instrument cluster fails to communicate with the ECM and TCM, the MIL is continuously lit by default. Under normal conditions, when the ignition switch is turned on the lamp remains lit for 15 seconds for a bulb check. Afterward, the MIL will light only for emissions related concerns. Whenever an engine misfire severe enough to damage the catalytic converter is detected, the MIL will flash on and off.

- ### CAMSHAFT POSITION ACTUATOR CONTROL SOLENOIDS

  A pair of duty cycle controlled solenoid valves that increase or decrease timing advance of the intake camshafts by controlling engine oil flow to the camshaft position actuators. When the duty cycle is greater than 50%, the oil flow from the solenoid causes the actuator to advance the camshaft position. When the duty cycle is less than 50%, the oil flow from the solenoid causes the actuator to retard the camshaft position. When the ECM determines that the desired camshaft position has been achieved, the duty cycle is commanded to 50% to hold the actuator so that the adjusted camshaft position is maintained. The solenoid winding resistance spec is 12 ± 2 ohms.

- ### EXHAUST GAS RECIRCULATION (EGR) VALVE

  A duty cycle controlled solenoid that, when energized, lifts the spring-loaded EGR valve pintle to open the valve. A value of 0% indicates an ECM command to fully close the EGR valve, and a value of 100% indicates an ECM command to fully open the EGR valve. The solenoid is enabled when the engine coolant temperature reaches 150°F (66°C) and the throttle is not closed or wide open. The solenoid winding resistance spec is 12 ± 2 ohms.

• FUEL INJECTORS

Electro-mechanical devices used to deliver fuel to the intake manifold at each cylinder. Each injector is individually energized once per camshaft revolution, in time with its cylinder's intake stroke. The injector winding resistance spec is 12 ± 2 ohms.

• IGNITION COILS

These six coils, mounted above the spark plugs, generate a high voltage to create a spark at each cylinder individually. Timing and dwell are controlled by the ECM. The coil primary resistance spec is 1 ± .5 ohms. The coil secondary resistance spec is 10K ± 2K ohms.

• GENERATOR FIELD

The ECM supplies this variable-duty cycle signal to ground the field winding of the generator (alternator), without the use of a separate voltage regulator. Increasing the duty cycle results in a higher field current and greater generator (alternator) output.

• EVAPORATIVE EMISSION (EVAP) CANISTER PURGE SOLENOID

A duty cycle controlled solenoid that regulates the flow of vapors stored in the canister to the intake manifold. The solenoid is enabled when the engine coolant temperature reaches 150°F (66°C). A duty cycle of 0% blocks vapor flow, and a duty cycle of 100% allows maximum vapor flow. The duty cycle is determined by the ECM, based on engine speed and load. The solenoid is also used for OBD II testing of the evaporative emission (EVAP) system. The solenoid winding resistance spec is 48 ± 6 ohms. There is also a service port with a schrader valve and cap installed on the hose between the purge solenoid and the canister.

• EVAPORATIVE EMISSION (EVAP) CANISTER VENT SOLENOID

When energized, the fresh air supply hose to the canister is blocked. The solenoid is only energized for OBD II testing of the evaporative emission (EVAP) system. The solenoid winding resistance spec is 48 ± 6 ohms.

• TORQUE CONVERTER CLUTCH (TCC) SOLENOID VALVE

A duty cycle controlled solenoid valve that applies the torque converter clutch by redirecting hydraulic pressure in the transaxle. With a duty cycle of 0%, the TCC is released. When torque converter clutch application is desired, the pulse width is increased until the clutch is fully applied. The solenoid will then maintain a 100% duty cycle until clutch disengagement is commanded. Then the pulse width is decreased back to 0%. If the brake pedal position switch closes, the duty cycle is cut to 0% immediately. The solenoid is enabled when the engine coolant temperature reaches 150°F (66°C), the brake switch is open, the transmission is in 3rd or 4th gear, and the vehicle is at cruise (steady throttle) above 40 mph. In addition, whenever the transmission fluid temperature is 248°F (120°C) or more, the TCM will command TCC lockup. The solenoid winding resistance spec is 12 ± 2 ohms.

• TRANSMISSION PRESSURE CONTROL (PC) SOLENOID

This duty cycle controlled solenoid controls fluid in the transmission valve body that is routed to the pressure regulator valve. By varying the duty cycle of the solenoid, the TCM can vary the line pressure of the transmission to control shift feel and slippage. When the duty cycle is minimum (10%), the line pressure will be maximized. When the duty cycle is maximum (90%), the line pressure will be minimized. The solenoid winding resistance spec is 6 ± 1 ohms.

- **TRANSMISSION SHIFT SOLENOIDS (SS 1 AND SS 2)**

  These solenoids control fluid in the transmission valve body that is routed to the 1-2, 2-3, and 3-4 shift valves. By energizing or de-energizing the solenoids, the TCM can enable a gear change. The solenoid winding resistance spec is 24 ± 4 ohms.

Gear	SS 1	SS 2
P, N, R, or 1	On	Off
2	Off	Off
3	Off	On
4 (OD)	On	On

SFI SYSTEM OPERATION AND COMPONENT FUNCTIONS

- **STARTING MODE**

  When the ignition switch is turned on, the ECM energizes the fuel pump relay for 2 seconds, allowing the fuel pump to build up pressure in the fuel system. Unless the engine is cranked within this two-second period, the fuel pump relay is de-energized to turn off the pump. The fuel pump relay will remain energized as long as the engine speed (CKP) signal to the ECM is 100 rpm or more.

- **CLEAR FLOOD MODE**

  When the accelerator pedal is fully depressed (pedal position of 80% or greater) and the engine speed is below 400 rpm, the ECM turns off the fuel injectors.

- **RUN MODE: OPEN AND CLOSED LOOP**

  - **OPEN LOOP**

    When the engine is first started and running above 400 rpm, the system operates in open loop. In open loop, the ECM does not use the oxygen sensor signals. Instead, it calculates the fuel injector pulse width from the throttle position sensors, the coolant and intake air temperature sensors, the MAF sensor, and the CKP sensor.

    The system will stay in open loop until all of these conditions are met:
    - Both upstream heated oxygen sensors (HO2S 1/1 and HO2S 2/1) are sending varying signals to the ECM.
    - The engine coolant temperature is above 68°F (20°C).
    - Ten seconds have elapsed since start-up.
    - Throttle position is less than 80%.

• CLOSED LOOP

When the oxygen sensor, engine coolant temperature sensor, and time conditions are met, and the throttle opening is less than 80%, the system goes into closed loop. Closed loop means that the ECM adjusts the fuel injector pulse widths for Bank 1 and Bank 2 based on the varying voltage signals from the upstream oxygen sensors. An oxygen sensor signal below 0.45 volts causes the ECM to increase injector pulse width. When the oxygen sensor signal rises above 0.45 volts in response to the richer mixture, the ECM reduces injector pulse width. This feedback trims the fuel control program that is based on the other sensor signals.

• ACCELERATION ENRICHMENT MODE

During acceleration, the ECM uses the increase in mass airflow and the rate of change in throttle position to calculate increased fuel injector pulse width. During wide open throttle operation, the control system goes into open loop mode.

• DECELERATION ENLEANMENT MODE

During deceleration, the ECM uses the decrease in mass airflow, the vehicle speed value, and the rate of change in throttle position to calculate decreased fuel injector pulse width.

• FUEL CUT-OFF MODE

The ECM will turn off the fuel injectors, for safety reasons, when the vehicle speed reaches 110 mph, or if the engine speed exceeds 6000 rpm.

OBD II SYSTEM OPERATION

• COMPREHENSIVE COMPONENT MONITOR

The OBD II diagnostic system continuously monitors all engine and transmission sensors and actuators for shorts, opens, and out-of-range values, as well as values that do not logically fit with other powertrain data (rationality). On the first trip during which the Comprehensive Component Monitor detects a failure that will result in emissions exceeding a predetermined level, the ECM or TCM will store a diagnostic trouble code (DTC), illuminate the malfunction indicator lamp (MIL), and store a freeze frame.

• SYSTEM MONITORS

The OBD II diagnostic system also actively tests some systems for proper operation while the vehicle is being driven. Fuel control and engine misfire are checked continuously. Oxygen sensor response, oxygen sensor heater operation, catalyst efficiency, EGR operation, EVAP integrity, variable valve timing, and thermostat operation are tested once or more per trip. When any of the System Monitors detects a failure that will result in emissions exceeding a predetermined level on two consecutive trips, the ECM will store a diagnostic trouble code (DTC) and illuminate the malfunction indicator lamp (MIL). Freeze frame data captured during the first of the two consecutive failures is also stored.

• **Fuel Control** - This monitor will set a DTC if the system fails to enter Closed Loop mode within 5 minutes of startup, or the Long Term Fuel Trim is excessively high or low anytime after the engine is warmed up, indicating the loss of fuel control. This is always the case when the Long Term Fuel Trim reaches its limit (+30% or -30%).

• **Engine Misfire** - This monitor uses the CKP sensor signal to continuously detect engine misfires, both severe and non-severe. If the misfire is severe enough to cause catalytic converter damage, the MIL will blink as long as the severe misfire is detected.

- **Oxygen Sensors** - This monitor checks the maximum and minimum output voltage, as well as switching and response times for all oxygen sensors. If an oxygen sensor signal remains too low or too high, switches too slowly, or not at all, a DTC is set.
- **Oxygen Sensor Heaters** - This monitor checks the current flow through each oxygen sensor heater. If the current flow is too high or too low, a DTC is set. Battery voltage is continuously supplied to the oxygen sensor heaters whenever the ignition switch is on. The heaters are grounded through the ECM.
- **Catalytic Converter** - This monitor compares the signals of the two upstream heated oxygen sensors to the signal from the downstream heated oxygen sensor to determine the ability of catalyst to store free oxygen. If the catalyst's oxygen storage capacity is sufficiently degraded, a DTC is set. This monitor will only run after the oxygen sensor heater and oxygen sensor monitors have run and passed for all three oxygen sensors.
- **EGR System** - This monitor uses the MAP sensor signal to detect changes in intake manifold pressure as the EGR valve is commanded open and closed. If the pressure changes too little or too much, a DTC is set.
- **EVAP System** - This monitor first turns on the EVAP vent solenoid to block the fresh air supply to the EVAP canister. Next, the EVAP purge solenoid is turned on to draw a slight vacuum on the entire EVAP system, including the fuel tank. Then the EVAP purge solenoid is turned off to seal the system. The monitor uses the Fuel Tank (EVAP) Pressure Sensor signal to determine if the EVAP system has any leaks. After the leak testing is completed, the EVAP vent solenoid is turned off to relieve the vacuum. If sufficient vacuum is not created, or decays too rapidly, or does not decay quickly at the conclusion of the leak test, a DTC is set. In order to run this monitor, the engine must be cold (below 86°F / 30°C) and the fuel level must be between ¼ and ¾ full.
- **Variable Valve Timing** - This monitor compares the desired valve timing with the actual timing indicated by the CMP sensors. If the timing is in error, or takes too long to reach the desired value, a DTC is set.
- **Engine Thermostat** - This monitor confirms that the engine warms up fully within a reasonable amount of time. If the coolant temperature remains too low for too long, a DTC is set.

- ## MONITOR READINESS STATUS

  The monitor readiness status indicates whether or not a particular OBD II diagnostic monitor has been run since the last time that DTCs were cleared from ECM and TCM memory. If the monitor has not yet run, the status will display on the Scan Tool as "Not Complete". If the monitor has been run, the status will display on the scan tool as "Complete". This does not mean that no faults were found, only that the diagnostic monitor has been run. Whenever DTCs are cleared from memory or the battery is disconnected, all monitor readiness status indicators are reset to "Not Complete". Monitor readiness status indicators are not needed for the Comprehensive Component, Fuel Control, and Engine Misfire monitors because they run continuously. The readiness status of the following system monitors can be read on the scan tool:

Oxygen Sensors	Oxygen Sensor Heaters	Catalytic Converter	EGR System
EVAP System	Variable Valve Timing	Engine Thermostat	

● WARM UP CYCLE

Warm Up Cycles are used by the ECM and TCM for automatic clearing of DTCs and Freeze Frame data as described below. To complete one warm up cycle, the engine coolant temperature must rise at least 40°F (22°C) and reach a minimum of 160°F (71°C).

● TRIP

A trip is a key-on cycle in which all enable criteria for a particular diagnostic monitor are met and the diagnostic monitor is run. The trip is completed when the ignition switch is turned off.

● DRIVE CYCLE

Most OBD II diagnostic monitors will run at some time during normal operation of the vehicle. However, to satisfy all of the different Trip enable criteria and run all of the OBD II diagnostic monitors, the vehicle must be driven under a variety of conditions. The following drive cycle will allow all monitors to run on this vehicle.

1. Ensure that the fuel tank is between ¼ and ¾ full.
2. Start cold (below 86°F / 30°C) and warm up until engine coolant temperature is at least 160°F (71°C).
3. Accelerate to 40-55 mph at 25% throttle and maintain speed for five minutes.
4. Decelerate without using the brake (coast down) to 20 mph or less, and then stop the vehicle. Allow the engine to idle for 10 seconds, turn the key off, and wait one minute.
5. Restart and accelerate to 40-55 mph at 25% throttle and maintain speed for two minutes.
6. Decelerate without using the brake (coast down) to 20 mph or less, and then stop the vehicle. Allow the engine to idle for 10 seconds, turn the key off, and wait one minute.

● FREEZE FRAME DATA

A Freeze Frame is a miniature "snapshot" (one frame of data) that is automatically stored in the ECM/TCM memory when an emissions-related DTC is first stored. If a DTC for fuel control or engine misfire is stored at a later time, the newest data are stored and the earlier data is lost. All parameter ID (PID) values listed under "Scan Tool Data" are stored in freeze frame data. The ECM/TCM stores only one single freeze frame record.

● STORING AND CLEARING DTCS & FREEZE FRAME DATA, TURNING THE MIL ON & OFF

● **One Trip Monitors**: A failure on the first trip of a "one trip" emissions diagnostic monitor causes the ECM or TCM to immediately store a confirmed DTC and Freeze Frame data, and turn on the MIL. All Comprehensive Component Monitor faults set a confirmed DTC on one trip.

● **Two Trip Monitors**: A failure on the first trip of a "two trip" emissions diagnostic monitor causes the ECM to store a pending DTC and Freeze Frame data. For all monitors, if the failure recurs on the next trip during which the monitor runs, regardless of the engine conditions, the ECM will store a confirmed DTC and turn on the MIL. In addition, for the misfire and fuel control monitors, if the failure recurs on the next trip during which the monitor runs and where conditions are similar to those experienced when the fault first occurred (engine speed within 375 rpm, engine load within 20%, and same hot/cold warm-up status), the ECM will store a confirmed DTC and turn on the MIL. If the second failure does not recur as described above, the pending DTC and Freeze Frame data are cleared from memory. All of the System Monitors are two trip monitors. Engine misfire which is severe enough to damage the catalytic converter is a two trip monitor. However, the MIL will always blink when the severe misfire is occurring.

● **Automatic Clearing**: When the vehicle completes three consecutive good/passing trips (three consecutive trips in which the monitor that set the DTC is run and passes), the MIL will be turned off, but the confirmed DTC and Freeze Frame will remain stored in ECM/TCM memory. For misfire and fuel control monitor faults, the three consecutive good/passing trips must take place under similar engine conditions (engine speed, load, and warm up condition) as the initial fault for the MIL to be turned off. If the vehicle completes 40 Warm Up cycles without the same fault recurring, the DTC and Freeze Frame are automatically cleared from the ECM/TCM memory.

● **Manual Clearing**: Any stored DTCs and Freeze Frame data can be erased using the scan tool, and the MIL (if lit) will be turned off. Although it is not the recommended method, DTCs and Freeze Frame data will also be cleared if the B+ power supply for the ECM/TCM is lost, or the battery is disconnected.

● **SCAN TOOL DATA**

These are different values and the minimum-to-maximum ranges for each data parameter that the OBD II scan tool is capable of displaying:

ECT: 248 to -40°F / 120 to -40°C / 0.0 to 5.0 v.
IAT: 248 to -40°F / 120 to -40°C / 0.0 to 5.0 v.
MAP/BARO: 20 to 101 kPa pressure /
    24 to 0 in.Hg. vacuum / 0.0 to 5.0 v.
MAF: 0 to 175 gm/sec / 0.0 to 5.0 volts
TP 1: 0 to 100% / 0.0 to 5.0 v.
TP 2: 0 to 100% / 0.0 to 5.0 v.
APP 1: 0 to 100% / 0.0 to 5.0 v.
APP 2: 0 to 100% / 0.0 to 5.0 v.
CKP (engine speed): 0 to 6000 rpm
Calculated Load Value: 0 to 100%
HO2S 1/1: 0.00 to 2.00 v.
HO2S 2/1: 0.00 to 2.00 v.
HO2S 1/2: 0.00 to 2.00 v.
Loop: Open / Closed
Valid Ignition Key: Yes / No
Fuel Enable: Yes / No
Bank 1 Injector Pulse Width: 0 to 15 ms
Bank 2 Injector Pulse Width: 0 to 15 ms
Bank 1 Long Term Fuel Trim: -30% to +30%
Bank 1 Short Term Fuel Trim: -30% to +30%
Bank 2 Long Term Fuel Trim: -30% to +30%
Bank 2 Short Term Fuel Trim: -30% to +30%
Ignition Timing Advance: 0 to 60° BTDC
Knock Sensor knock detected: Yes / No
Throttle Actuator Control: 0 to 100%
Battery Voltage: 0 to 18 v.
Generator Field: 0 to 100%
Intake Cam 1 Desired Advance: 0 to 40°
Intake Cam 2 Desired Advance: 0 to 40°
CMP 1: 0 to 40°
CMP 2: 0 to 40°
EGR Valve Opening Desired: 0 to 100%
EGR Position Sensor: 0 to 100% / 0.0 to 5.0 v.

Evap Purge Solenoid: 0 to 100%
Evap Vent Solenoid: On / Off
Fuel Tank (EVAP) Pressure: -14.0 to +14.0
    in.H₂O / -0.5 psi to 0.5 psi / 0.0 to 5.0 v.
Fuel Tank Level: 0 to 100% / 0.0 to 5.0 v.
P/S Switch: On / Off
Brake Switch: On / Off
A/C Request: On / Off
A/C Pressure: 25 to 450 psi / 0.0 to 5.0 v.
A/C Clutch: On / Off
Fan Control: On / Off
Fuel Pump: On / Off
TR: P/N, R, 1, 2, 3, OD
TFT: 248 to -40°F / 120 to -40°C / 0.0 to 5.0 v.
VSS: 0 to 110 mph
TSS: 0 to 6000 rpm
SS 1: On / Off
SS 2: On / Off
TCC: 0 to 100%
PC: 0 to 100%
MIL: On / Off / Flashing
Confirmed DTCs: P####, U####, etc.
Pending DTCs: P####, U####, etc.
Monitor Status for this trip: Disabled /
    Not Complete / Complete
Time elapsed since engine start: hh:mm:ss
Distance traveled with MIL on: #### miles/km
Distance traveled since DTCs cleared:
    #### miles/km
Number of warm-up cycles since DTCs
    cleared: ###
Software Calibration ID # (CAL ID)
Software Verification Number (CVN)
Vehicle Identification Number (VIN)

Temperature °F	Temperature °C	Sensor Voltage
248	120	0.25
212	100	0.46
176	80	0.84
150	66	1.34
140	60	1.55
104	40	2.27
86	30	2.60
68	20	2.93
32	0	3.59
-4	-20	4.24
-40	-40	4.90

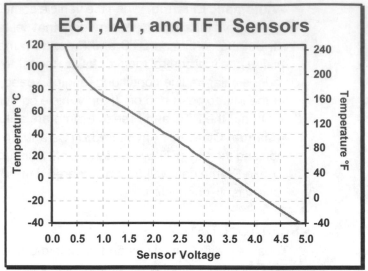

Vacuum at seal level (in. Hg.)	Manifold Absolute Press. (kPa)	Sensor Voltage
0	101.3	4.50
3	91.2	4.00
6	81.0	3.50
9	70.8	3.00
12	60.7	2.50
15	50.5	2.00
18	40.4	1.50
21	30.2	1.00
24	20.1	0.50

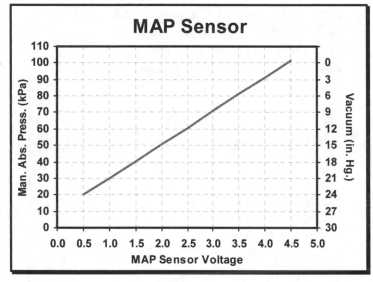

Mass Airflow (gm/sec)	Sensor Voltage
0	0.20
2	0.70
4	1.00
8	1.50
15	2.00
30	2.50
50	3.00
80	3.50
110	4.00
150	4.50
175	4.80

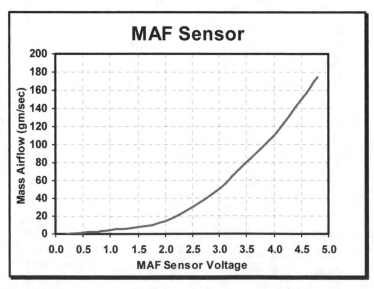

Throttle Position (% open)	TP 1 Sensor Voltage	TP 2 Sensor Voltage
0	4.50	0.50
5	4.30	0.70
10	4.10	0.90
15	3.90	1.10
20	3.70	1.30
25	3.50	1.50
40	2.90	2.10
50	2.50	2.50
60	2.10	2.90
75	1.50	3.50
80	1.30	3.70
100	0.50	4.50

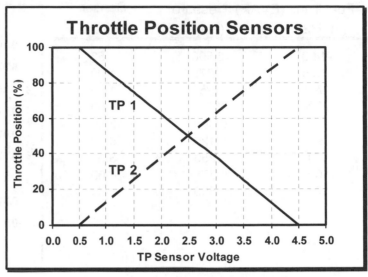

Accelerator Pedal Pos'n (% depressed)	APP 1 Sensor Voltage	APP 2 Sensor Voltage
0	0.50	1.50
5	0.65	1.65
10	0.80	1.80
15	0.95	1.95
20	1.10	2.10
25	1.25	2.25
40	1.70	2.70
50	2.00	3.00
60	2.30	3.30
75	2.75	3.75
80	2.90	3.90
100	3.50	4.50

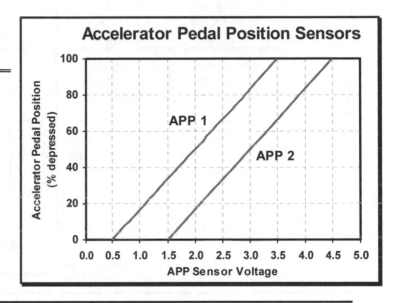

A/C High Side Pressure (psi)	Sensor Voltage
25	0.25
50	0.50
100	1.00
150	1.50
200	2.00
250	2.50
300	3.00
350	3.50
400	4.00
450	4.50

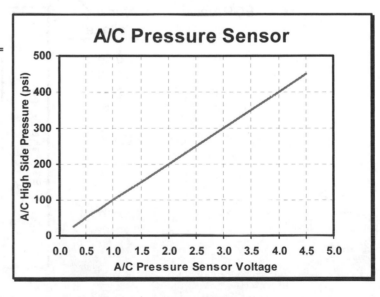

Fuel Tank (EVAP) Pressure		Sensor
(in. H2O)	(psi)	Voltage
-14.0	-0.500	0.50
-10.5	-0.375	1.00
-7.0	-0.250	1.50
-3.5	-0.125	2.00
0.0	0.000	2.50
3.5	0.125	3.00
7.0	0.250	3.50
10.5	0.375	4.00
14.0	0.500	4.50

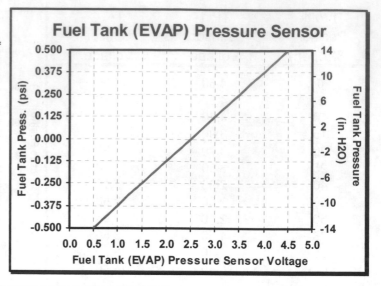

Fuel Level	Sensor
(% full)	Voltage
0	0.50
25	1.50
50	2.50
75	3.50
100	4.50

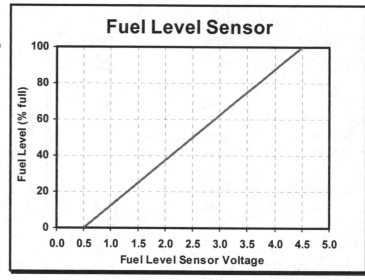

EGR Valve	Sensor
(% open)	Voltage
0	0.50
25	1.50
50	2.50
75	3.50
100	4.50

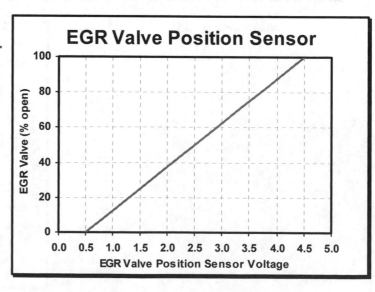

↑ To B+ in Start and Run Fuse #2

↑ To B+ at all times Fuse #3

↑ To B+ in Start and Run Fuse #4

ECM

1 +5 v.
2 Ign.
3 B+

18 TP 1 — Throttle Position Sensors — c, a, b, d
19 TP 2

20 EGR Position — EGR Valve Position Sensor — a, c, b

21 CKP + — Crankshaft Position Sensor — a, b
22 CKP −

23 MAP — MAP Sensor — a, c, b

24 KS — Knock Sensor — b, a — →+5v

25 ECT — ECT Sensor — b, a — →+5v

26 IAT — IAT Sensor — b, a — →+5v

27 A/C Pressure — A/C Pressure Sensor — a, c, b

28 Fuel Tank Pressure — Fuel Tank Pressure Sensor — a, c, b

29 Fuel Level — Fuel Level Sensor — a, c, b

31 Sensor Ground

Fuel Pump Relay — FP 4 — c, d, b, a — Fuel Pump — a, b

A/C Clutch Relay — A/C Clutch 5 — c, d, b, a — A/C Clutch — a, b

Fan Control Relay — Fan Control 6 — c, d, b, a — Cooling Fan Motor — a, b

Coil 1 7 — c, b, a
Coil 3 8 — c, b, a
Coil 5 9 — c, b, a — Bank 1 — 5, 3, 1

Coil 2 10 — c, b, a
Coil 4 11 — c, b, a
Coil 6 12 — c, b, a — Bank 2 — 6, 4, 2

B+ B+ — TAC 14 — a — Throttle Actuator Control Motor
TAC 15 — b

Cam 1 16 — Cam 1 Pos'n Solenoid — b, a
Cam 2 17 — Cam 2 Pos'n Solenoid — b, a — → A — see Diagram 2

Ground
32

see Diagram 2 — B

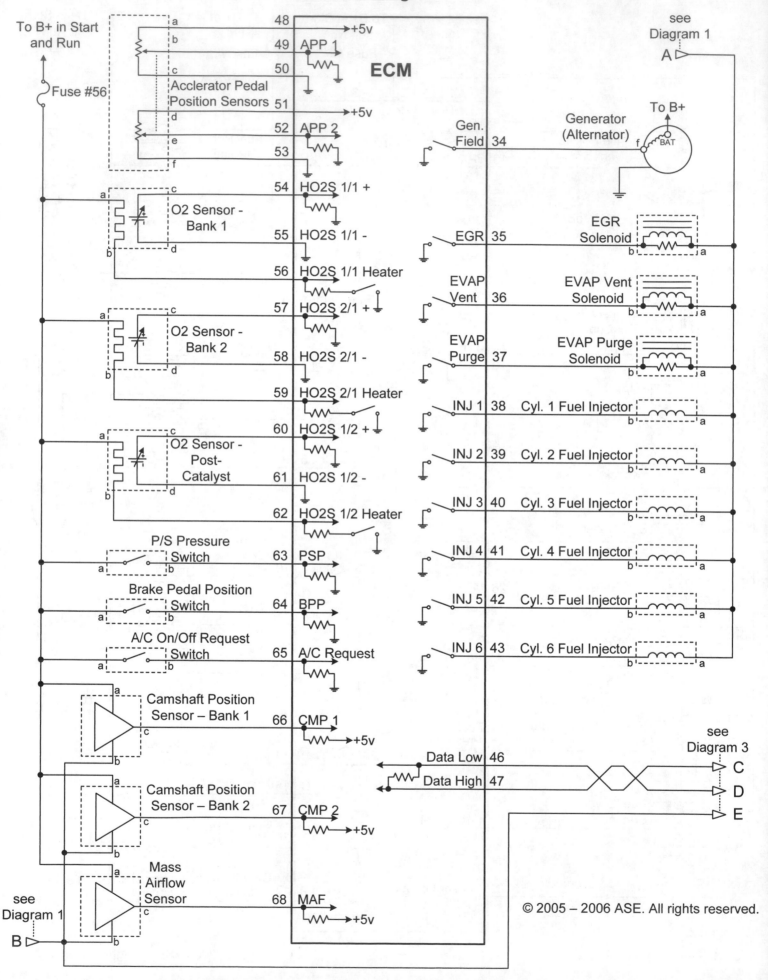

To B+ in Start and Run Fuse #69

To B+ at all times Fuse #70

To B+ in Start and Run Fuse #72

69 Ign.

70 B+

TCM

Trans. Fluid Temp. Sensor

80 TFT →+5v

b a

81 VSS +

82 VSS −

Vehicle Speed Sensor

a

b

83 TSS +

84 TSS −

Trans. Turbine Shaft Speed Sensor

a

b

TCC Solenoid

TCC 72

b a

SS 1 74

Trans. Shift Solenoid 1

b a

SS 2 75

Trans. Shift Solenoid 2

b a

PC 76

Trans. Pressure Control Solenoid

b a

85 Park/Neutral

86 Rev

87 1st

88 2nd

89 3rd

90 OD

Data High 78

Data Low 79

b

c

d

e

f

a

g

Transmission Range Switch

91 Sensor Ground

Ground

92

see Diagram 2

C

D

E

To B+ in Start and Run Fuse #99

93 Ign.

96 Ant. +

Data Low 94

Data High 95

97 Ant. −

Ground

98

IMMOBILIZER MODULE

Transponder Key

To B+ at all times

Fuse #74

99 Ign.

100 Data Low

101 Data High

Tachometer

Speedometer

Fuel Level

Temperature

Ign.

MIL

Immobilizer

Ground

102

1	2	3	4	5	6	7	8
9	10	11	12	13	14	15	16

DATA LINK CONNECTOR

INSTRUMENT CLUSTER

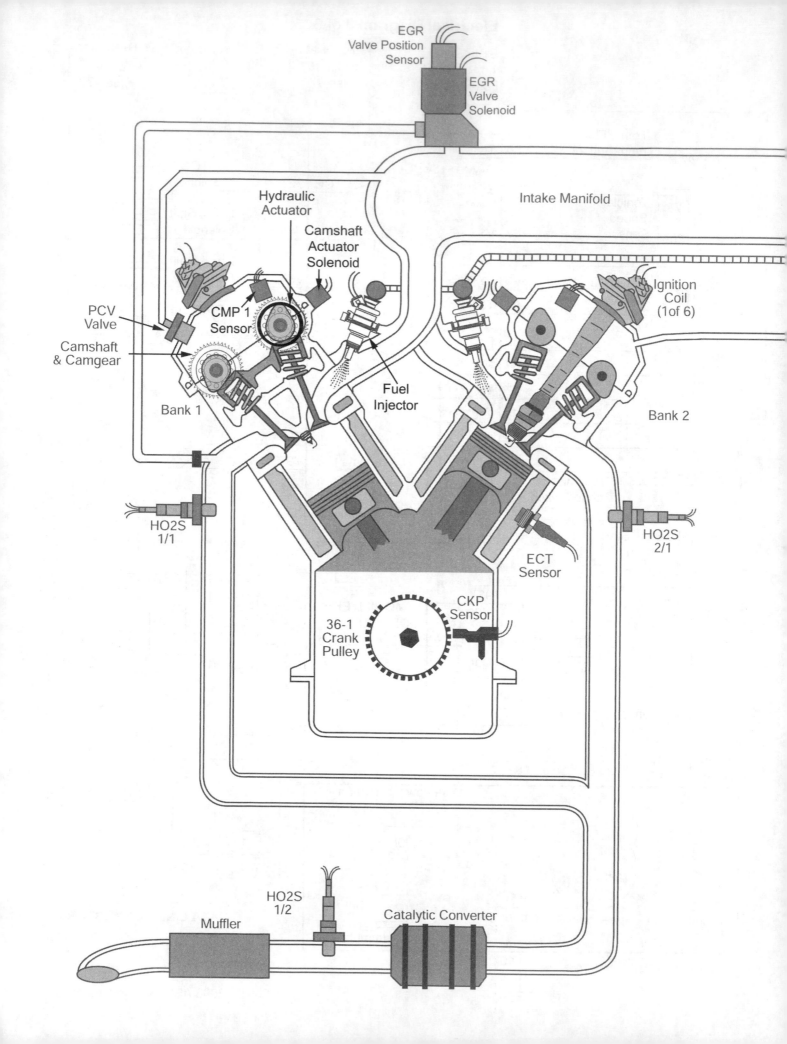

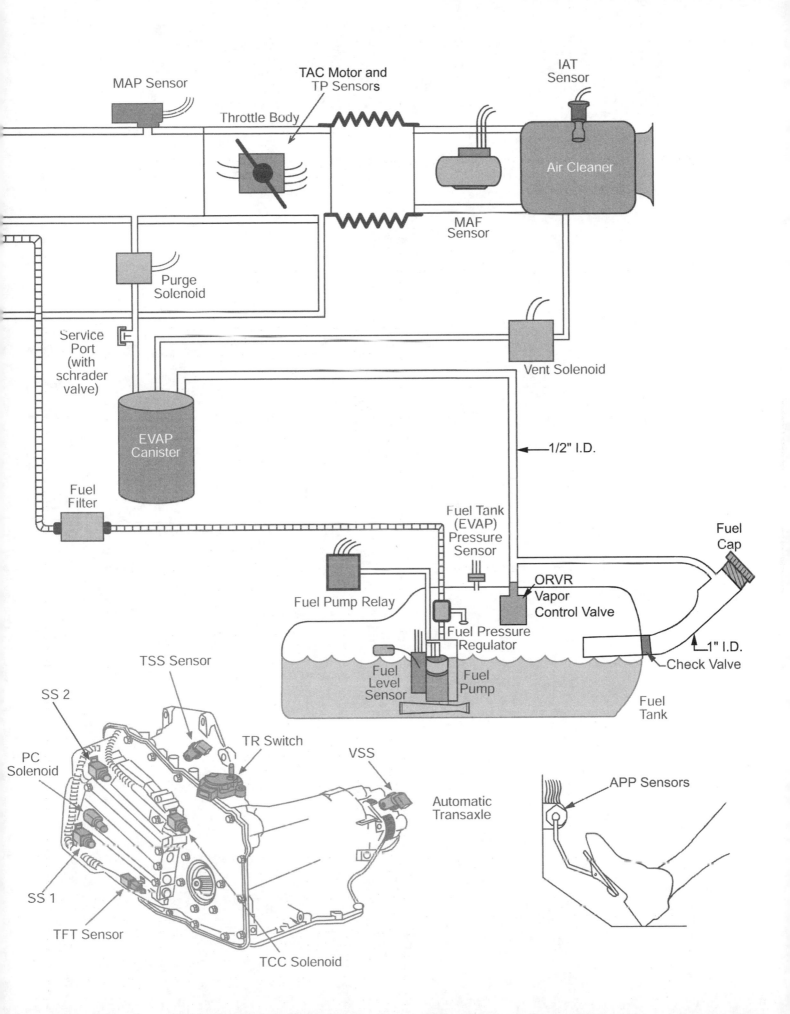

MAP Sensor

TAC Motor and
TP Sensors

Throttle Body

IAT
Sensor

Air Cleaner

MAF
Sensor

Purge
Solenoid

Vent Solenoid

Service
Port
(with
schrader
valve)

1/2" I.D.

EVAP
Canister

Fuel
Filter

Fuel Tank
(EVAP)
Pressure
Sensor

Fuel
Cap

Fuel Pump Relay

ORVR
Vapor
Control Valve

1" I.D.

Fuel Pressure
Regulator

Check Valve

Fuel
Level
Sensor

Fuel
Pump

Fuel
Tank

TSS Sensor

SS 2

TR Switch

VSS

PC
Solenoid

Automatic
Transaxle

APP Sensors

SS 1

TFT Sensor

TCC Solenoid

5 | Sample Test for Practice

Sample Test

Please note the letter and number in parentheses following each question. They match the task in Section 4 that discusses the relevant subject matter. You may want to refer to the overview using the cross-referencing key to help with questions posing problems for you.

1. A vehicle has been repaired that had a broken hose going to the MAP sensor. Technician A says make sure all old codes are erased. Technician B says the vehicle should be driven to verify the repair fixed the vehicle. Who is correct?
   A. A only
   B. B only
   C. Both A and B
   D. Neither A nor B   (B.23)

SCAN TOOL DATA			
Engine Coolant Temperature (ECT) Sensor 210°F/ 99°C/ .50 volts	Intake Air Temperature (IAT) Sensor 96°F/ 35°C/ 2.4volts	Mass Airflow Sensor (MAF) 50gm/sec/3.0 volts	Throttle Actuator Control Motor (TAC) 35 percent
Throttle Position Sensor 1 (TP1) 0 percent / 0.00 volts	Throttle Position Sensor 2 (TP2) 35 percent / 2.00 volts	Accelerator Pedal Position Sensor 1 (APP1) 100 percent / 3.5 volts	Accelerator Pedal Position Sensor 2 (APP2) 100 percent / 4.5 volts
Crankshaft Position Sensor (CPS) 1900 rpm	Heated Oxygen Sensor Bank 1 (HO2S1/1) .01 - .90 volts	Heated Oxygen Sensor Bank 2 (HO2S2/1) 0.1 - .90 volts	Heated Oxygen Sensor Post-Cat (HO2S1/2) 0.4 volts
Vehicle Speed Sensor (VSS) 55 MPH	EVAP Canister Purge Solenoid 30 percent	EVAP Canister Vent Solenoid OFF	Fuel Pump Relay (FP) ON
Measured Ignition Timing °BTDC	Base Timing 10°	Actual Timing 26°	

2. The composite vehicle is sluggish on acceleration and will not exceed 55 MPH and the MIL is on. A check of basic systems shows all components to be in good condition. The scan tool data in the table above was obtained on a test drive. The throttle pedal was wide open going 55 MPH. What is limiting the vehicle speed?
   A. An incorrect Mass Airflow Sensor reading.
   B. The vehicle speed sensor (VSS) is faulty.
   C. The Throttle Actuator Control Motor is stuck at 35% opening.
   D. The Throttle Position Sensor 1 is inaccurate.   (B.2)

SCAN TOOL DATA			
Engine Coolant Temperature (ECT) Sensor 210°F/ 99°C/ .50 volts	Intake Air Temperature (IAT) Sensor 145°F/ 63°C/ 1.45 volts	Mass Airflow Sensor (MAF) 2.0 gm/sec/0.7 volts	Throttle Actuator Control Motor (TAC) 5 percent
Throttle Position Sensor 1 (TP1) 5 percent / 4.30 volts	Throttle Position Sensor 2 (TP2) 5 percent / 0.70 volts	Accelerator Pedal Position Sensor 1 (APP1) 5 percent / 0.65 volts	Accelerator Pedal Position Sensor 2 (APP2) 5 percent / 1.65 volts
Crankshaft Position Sensor (CPS) 850 rpm	Heated Oxygen Sensor Bank 1 (HO2S 1/1) .01 - .90 volts	Heated Oxygen Sensor Bank 2 (HO2S 2/1) 0.1 - .90 volts	Heated Oxygen Sensor Post-Cat (HO2S 1/2) 0.4 volts
Battery Voltage (B+) 9.4 volts	EVAP Canister Purge Solenoid 0 percent	EVAP Canister Vent Solenoid OFF	Fuel Pump Relay (FP) ON
Vehicle Speed Sensor (VSS) 0 MPH	Open/Closed Loop CLOSED	Malfunction Indicator Lamp (MIL) OFF	Ignition Timing Advance 2 BTDC
Measured Ignition Timing °BTDC	Base Timing 10°	Actual Timing 12°	

3. The data shown is for a composite vehicle that is exhibiting slow cranking, a hunting idle, and decreased power. What is the MOST likely cause for this condition?
    A. Reference voltage not stable
    B. Oxygen sensor contaminated
    C. Insufficient output from the charging system
    D. A faulty starter   (A.4 and A.5)

SCAN TOOL DATA			
Engine Coolant Temp. Sensor (ECT) 215°F/ 102°C / .42v.	Intake Air Temperature Sensor (IAT) 107°F/ 42°C / 2.24v.	Mass Airflow Sensor (MAF) .0 gm/sec/ 0.20v.	Throttle Actuator Control Motor (TAC) 5 percent
Throttle Position Sensor 1 (TP 1) 5 percent / 4.30 v.	Throttle Position Sensor 2 (TP2) 5 percent / 0.70 v.	Accelerator Pedal Position Sensor 1 (APP1) 5 percent / 0.65 v.	Accelerator Pedal Position Sensor 2 (APP2) 5 percent / 1.65 v.
Crankshaft Position Sensor (CPS) 750 rpm	Heated Oxygen Sensor Bank 1 (HO2S 1/1) .01-.90 v.	Heated Oxygen Sensor Bank 2 (HO2S 2/1) 0.1-.90 v.	Heated Oxygen Sensor Post-Cat (HO2S 1/2) 0.4 v.
Battery Voltage (B+) 14.2 volts	EVAP Canister Purge Solenoid 0 percent	EVAP Canister Vent Solenoid OFF	Fuel Pump Relay (FP) ON
Vehicle Speed Sensor (VSS) 0 mph	Open/Closed Loop CLOSED	Malfunction Indicator Lamp (MIL) OFF	Ignition Timing Advance 1 BTDC

4. Refer to the scan data. The engine is difficult to start and is sluggish on acceleration. Emissions and engine vacuum are good. What is the MOST likely problem?
    A. The TAC valve is in the wrong position.
    B. The TPS 1 and TPS 2 are shorted.
    C. The MAF sensor is sending incorrect values.
    D. The fuel pressure is too high.   (B.8 and D.5)

5. A customer complains that his fuel-injected vehicle has poor fuel economy and power when the outside air is cold (below 30°F). During the summer, the vehicle performs better. After performing the basic checks on fuel pressure, engine mechanical conditions, and ignition system and finding that all systems are performing properly, which of the following should the technician check next?
   A. Check PCM power and grounds.
   B. Check fuel delivery volume.
   C. Test ignition coil output.
   D. Ensure that the coolant thermostat is not stuck open.   (A.8)

6. An SFI vehicle will not start. It is verified that there is no injector pulse or spark. Technician A says a faulty crankshaft position sensor could be the cause. Technician B says a throttle position sensor stuck at WOT could be the cause. Who is right?
   A. Technician A only
   B. Technician B only
   C. Both A and B
   D. Neither A nor B   (B.2, C.2, C.7, C.8 and C.12)

7. A vehicle has failed the I/M inspection. The vehicle has no live or history faults stored in the PCM. Technician A says to check for any technical service bulletins pertaining to the failure. Technician B says a good visual inspection should be done before any testing is performed. Who is correct?
   A. A only
   B. B only
   C. Both A and B
   D. Neither A nor B   (F.2)

8. A technician is diagnosing a suspected failed catalytic converter in a vehicle OBD II. The waveform of the downstream HO_2S is about the same as the upstream HO_2S waveform. What does this prove?
   A. The catalytic converter is working normally.
   B. The upstream sensor has failed.
   C. The converter has failed.
   D. Nothing, two waveforms should be the same.   (B.17 and E.8)

9. A vehicle exhibits a hard starting complaint, and a faulty ECT is suspected. Technician A says to check the Engine Coolant Temperature resistance and compare with the resistance value chart in the service manual. Technician B says to check for a five volt reference voltage from the PCM to ECT. Who is correct?
   A. A only
   B. B only
   C. Both A and B
   D. Neither A nor B   (B.15)

10. Two Technicians are discussing multiplexing. Technician A says that an SAE class A communications network is faster than an SAE class C communications network. Technician B says multiplexing eliminates redundant sensors and dedicated wiring for multiple sensors. Who is right?
    A. Technician A only
    B. Technician B only
    C. Both A and B
    D. Neither A nor B   (B.2)

11. A vehicle has repeated MIL illumination with no fault codes being set. Technician A says to use a scan tool and check for the software version of the PCM. Technician B says the PCM is reprogrammed through the Data Link Connector. Who is correct?
    A. A only
    B. B only
    C. Both A and B
    D. Neither A nor B   (B.5)

12. When using a digital storage oscilloscope (DSO) to test an O_2 sensor, it is important to confirm that the sensor is cycling within the proper range of rich/lean switching ratio or frequency (Hz). With the engine in closed loop at 2,500 rpm, what is the correct oxygen sensor frequency found in a properly operating system?
    A. 0 to 2 Hz
    B. 5 to 25 Hz
    C. 0.5 to 5 Hz
    D. 1 to 10 Hz   (B.2 and B.17)

13. A vehicle with an enhanced EVAP system has a small leak around the filler neck of the fuel tank. Technician A says this will cause the PCM to illuminate the MIL. Technician B says the MIL will illuminate if the vehicle is emitting emissions 1.5 times federal test procedure. Who is correct?
    A. A only
    B. B only
    C. Both A and B
    D. Neither A nor B   (F.12)

14. When discussing a vehicle with computerized engine controls, Technician A says that major components, such as a camshaft, can be replaced with aftermarket parts that have different specifications. Technician B says that emission-control devices can be bypassed for better performance. Who is right?
    A. Technician A only
    B. Technician B only
    C. Both A and B
    D. Neither A nor B   (A.1)

15. While discussing a 1995 vehicle with a no-start condition, Technician A says that the service manual for that year vehicle is not available. Technician A thinks that the 1996 model is about the same and that information can be used for diagnosis. Technician B says to properly diagnose the vehicle, the manual for the 1995 model will be needed for proper diagnostic procedures. Who is right?
    A. Technician A only
    B. Technician B only
    C. Both A and B
    D. Neither A nor B   (A.2 and A.3)

16. Technician A says an OBD II-compliant vehicle may check the EVAP system for purge flow during operation. Technician B says an OBD II-compliant vehicle may check the EVAP system for leaks. Who is right?
    A. Technician A only
    B. Technician B only
    C. Both A and B
    D. Neither A nor B   (E.3)

17. When discussing service procedures, Technician A says confirming the customer's complaint should be the first step. Technician B says locating relevant service information should be the first step. Who is right?
    A. Technician A only
    B. Technician B only
    C. Both A and B
    D. Neither A nor B   (A.2 and A.4)

18. A technician obtains a block-learn number (long-term fuel trim) of 250. What can be said of the operating condition of this vehicle?
    A. The PCM is correcting for an overly rich mixture.
    B. The PCM is correcting for an overly lean mixture.
    C. The ignition timing is too advanced.
    D. The ignition timing is too retarded.   (B.2 and B.3)

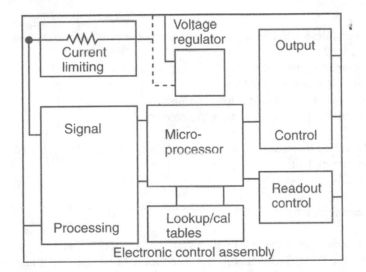

19. Refer to the figure. In which section of the electronic control assembly would you find the read-only memory (ROM)?
    A. Microprocessor
    B. Signal processor
    C. Readout control
    D. Lookup/calibration tables   (B.3)

20. A vehicle with OBD II has a fault stored in the PCM. After clearing the fault, the fault will not reappear. Technician A says to look up the enabling criteria for the fault in order to duplicate the conditions that set the fault. Technician B says the drive cycle for the faulty system in question should be researched. Who is correct?
    A. A only
    B. B only
    C. Both A and B
    D. Neither A nor B   (B.6)

21. Technician A says scan tools display diagnostic trouble codes (DTC). Technician B says that scan tools can be used to compare sensor parameters and make conclusions based on the scan tool data. Who is right?
    A. Technician A only
    B. Technician B only
    C. Both A and B
    D. Neither A nor B   (B.2, B.3 and B.8)

22. A Technician is diagnosing a vehicle with the MIL illuminated. Technician A says to check for freeze frame data stored in the PCM at the time of MIL illumination. Technician B says a Type A emission related code may cause the MIL to flash. Who is correct?
    A. A only
    B. B only
    C. Both A and B
    D. Neither A nor B   (B.7)

23. What is the most destructive result of disconnecting the battery from the vehicle while the engine is running?
    A. Deprogramming the radio
    B. System voltage exceeding 16 volts
    C. Erasing digital trip odometer information
    D. Erasing the adaptive strategy from the computer.   (B.9)

24. A rough idle caused by reduced airflow to one cylinder only could be caused by a:
    A. slipped timing belt.
    B. broken timing chain.
    C. leaking intake valve.
    D. worn camshaft lobe.   (A.6)

25. When trying to connect to the PCM with a scan tool, the technician only gets a no response message from the PCM. Technician A says a service manual and digital multimeter may be required to continue any further diagnosis. Technician B says the vehicle needs a new PCM. Who is correct?
    A. A only
    B. B only
    C. Both A and B
    D. Neither A nor B   (B.11)

26. A vehicle has a no start condition. The scan tool reveals a no fuel allowed message in the data stream. Technician A says the no start may be a fault with the anti-theft system. Technician B says the immobilizer in the key may have failed. Who is correct?
    A. A only
    B. B only
    C. Both A and B
    D. Neither A nor B   (B.12)

27. A voltage-drop test is performed on the fuel pump circuit. Technician A says a low-voltage drop across the fuel pump indicates that the fuel pump is serviceable. Technician B says that a high-voltage drop across the fuel pump relay switch indicates that the switch is serviceable. Who is right?
    A. Technician A only
    B. Technician B only
    C. Both A and B
    D. Neither A nor B   (B.13 and B.14)

28. Which of the following problems would have the greatest effect toward lowering CO_2 emissions?
    A. Excessive air inlet temperatures
    B. Lean air/fuel mixture
    C. Very retarded ignition timing
    D. Dirty fuel injectors   (F.4)

29. Engine Coolant Temperature integrity is being discussed. Technician A says the ECT readings can be compared with the IAT readings after the vehicle has set for at least eight hours. Technician B says the ECT should set a fault code if it is faulty. Who is correct?
    A. A only
    B. B only
    C. Both A and B
    D. Neither A nor B   (B.16)

Pre HO2S   Post HO2S

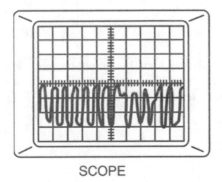   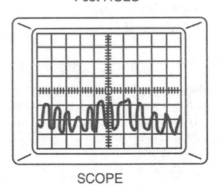

SCOPE   SCOPE

30. Refer to the figure and observe the O_2 signal waveforms. The comparison of the two waveforms indicates:
    A. a faulty pre-HO_2S sensor.
    B. a faulty post-HO_2S sensor.
    C. an overly rich mixture.
    D. a failing catalytic converter.   (B.17)

31. Which of the following conditions can result in elevated NO_x emissions?
    A. Excessively rich air/fuel mixture
    B. Lower-than-normal coolant temperature
    C. Carbon-buildup in the combustion chambers
    D. Restricted air filter   (A.9, A.12 and F.11)

SCAN TOOL DATA			
Engine Coolant Temperature (ECT) Sensor 68°F/ 20°C/ 2.93 volts	Intake Air Temperature (IAT) Sensor 68°F/ 20°C/ 2.93 volts	Mass Airflow Sensor (MAF) 2 gm/sec/0.70 volts	Throttle Actuator Control Motor (TAC) 0 percent
Throttle Position Sensor 1 (TP1) 0 percent / 4.50 volts	Throttle Position Sensor 2 (TP2) 0 percent / 50 volts	Accelerator Pedal Position Sensor 1 (APP1) 0 percent / .50 volts	Accelerator Pedal Position Sensor 2 (APP2) 0 percent / 1.50 volts
Crankshaft Position Sensor (CPS) 0 rpm	Heated Oxygen Sensor Bank 1 (HO2S 1/1) .01 volts	Heated Oxygen Sensor Bank 2 (HO2S 2/1) 0.1 volts	Heated Oxygen Sensor Post-Cat (HO2S 1/2) 0.0 volts
Battery Voltage (B+) 12.3 volts	EVAP Canister Purge Solenoid 0 percent	EVAP Canister Vent Solenoid OFF	Fuel Pump Relay (FP) OFF
Vehicle Speed Sensor (VSS) 0 MPH	Open/Closed Loop OPEN	Malfunction Indicator Lamp (MIL) OFF	Ignition Timing Advance 0 BTDC

Measured Ignition Timing °BTDC   Base Timing 10°   Actual Timing 11°

32. The composite vehicle will not start. The scan data above was taken while cranking the engine. Technician A says the scan data indicates a problem with the fuel pump relay. Technician B says the scan data indicates a problem with the crankshaft position sensor or circuit. Who is correct?
    A. Technician A only
    B. Technician B Only
    C. Both A and B
    D. Neither A nor B   (C.2 and C.8)

33. During a cylinder balance test, a cylinder's O_2 levels stay high, but do not change when the cylinder is shorted. Technician A says this could be caused by low compression in that cylinder. Technician B says this could be caused by a vacuum leak in that cylinder's runner. Who is right?
    A. Technician A only
    B. Technician B only
    C. Both A and B
    D. Neither A nor B   (A.7)

34. A vehicle has been towed to the repair shop from another repair shop with both the power-train control module and the transmission control module failed. Technician A says use of a test light for checking voltages at the module connectors could be the cause. Technician B says improper jump starting techniques could cause this. Who is right?
    A. Technician A only
    B. Technician B only
    C. Both A and B
    D. Neither A nor B   (A.13)

35. All of the following conditions could cause low fuel pressure in the composite vehicle **EXCEPT:**
    A. a restricted fuel filter.
    B. restricted fuel-pressure regulator.
    C. low voltage supply to the fuel pump.
    D. poor ground for the fuel pump.   (D.9)

36. A vehicle has failed an I/M test with the following readings:

	HC	CO	NO_x
Cutpoints	0.9	14	1.9
Actual Readings	1.0	21	1.8

The technician repaired the cause of the HC and CO failure. The following readings were obtained after the repair:
    The technician repaired the cause of the HC and CO failure. The following readings were obtained after the repair:

	HC	CO	NO_x
Cutpoints	0.9	14	1.9
Actual Readings	0.6	11	2.5

Why is the NO_x reading now higher than allowed?
- A. The heating effect of the lean air/fuel mixture was masking a NO_x failure.
- B. The cooling effect of the lean air/fuel mixture was masking a NO_x failure.
- C. The heating effect of a rich air/fuel mixture was masking a NO_x failure.
- D. The cooling effect of a rich air/fuel mixture was masking a NO_x failure.

(F.3, F.10, F.11 and F.16)

37. An O_2 sensor signal should transition from rich to lean in approximately:
- A. 200 ms
- B. 300 ms
- C. 100 ms
- D. 10 ms   (B.17)

38. A vehicle has a misfire under load and has high HC. The technician uses an oscilloscope and determines number two cylinder's firing voltage is 15 KV higher than the others. It also has a lower-than-normal power contribution from the same cylinder under load. Which of the following is MOST likely the cause of this problem?
- A. Faulty crankshaft position sensor
- B. An open number two spark plug wire
- C. Low compression in cylinder number two
- D. Shorter spark plug in number two cylinder   (C.9)

39. Technician A says that using a fluid other than the manufacturer's specified type will not have an effect on transmission performance. Technician B says that DEXRON-III® is a universal fluid that can be used in all transmissions. Who is right?
- A. Technician A only
- B. Technician B only
- C. Both A and B
- D. Neither A nor B   (A.10)

40. The PCM has a faulty injector driver circuit. Technician A says to replace the PCM. Technician B says to check injector resistance. Who is right?
- A. Technician A only
- B. Technician B only
- C. Both A and B
- D. Neither A nor B   (B.20)

SCAN TOOL DATA			
Engine Coolant Temperature (ECT) Sensor 176°F/ 80°C/ 0.84 volts	Intake Air Temperature (IAT) Sensor 86°F/ 30°C/ 2.60 volts	Mass Airflow Sensor (MAF) 110 gm/sec/4.00 volts	Throttle Actuator Control Motor (TAC) 80 percent
Throttle Position Sensor 1 (TP1) 80 percent / 1.30 volts	Throttle Position Sensor 2 (TP2) 80 percent / 3.70 volts	Accelerator Pedal Position Sensor 1 (APP1) 80 percent / 2.90 volts	Accelerator Pedal Position Sensor 2 (APP2) 80 percent / 3.90 volts
Crankshaft Position Sensor (CPS) 2000 rpm	Heated Oxygen Sensor Bank 1 (HO2S 1/1) .01-.90 volts	Heated Oxygen Sensor Bank 2 (HO2S 2/1) 0.1-.90 volts	Heated Oxygen Sensor Post-Cat (HO2S 1/2) 0.4 volts
Battery Voltage (B+) 16.8 volts	EVAP Canister Purge Solenoid 70 percent	EVAP Canister Vent Solenoid OFF	Fuel Pump Relay (FP) ON
Vehicle Speed Sensor (VSS) 60 MPH	Open/Closed Loop CLOSED	Malfunction Indicator Lamp (MIL) OFF	Ignition Timing Advance 12 BTDC

Measured Ignition Timing °BTDC   Base Timing 10°   Actual Timing 11°

41. The composite vehicle has experienced repeated electrical component failures. Using the data shown, what is the MOST likely cause of these failures?
    A. Overcharging generator
    B. Bad ground on the ECM
    C. Not enough data to determine cause   (B.22)
    D. Failed TPS sensors

42. Technician A says the I/M240 test can accurately measure HC and CO emissions prior to the converter. Technician B says the I/M240 test measures vehicle emissions in grams per mile. Who is right?
    A. Technician A only
    B. Technician B only
    C. Both A and B
    D. Neither A nor B   (F.3 and F.4)

43. Which term best describes the output signal of a hall-effect switch?
    A. Serial
    B. Analog
    C. Digital
    D. AC   (C.2 and C.3)

44. Which of the following sensors sends a crankshaft position signal to the PCM?
    A. CKP sensor
    B. TP sensor
    C. VS sensor
    D. CMP sensor   (C.3)

45. Technician A says a 12-volt test lamp can be used to check for voltage at the ignition primary coil winding. Technician B says a self-powered test lamp also can be used. Who is right?
    A. Technician A only
    B. Technician B only
    C. Both A and B
    D. Neither A nor B   (C.7, C.8 and C.12)

46. In what model year is full OBD II compliance mandated for passenger cars?
    A. 1994
    B. 1996
    C. 1997
    D. 1995   (B.2)

47. An engine will not start. With the engine cranking, a test lamp connected between the coil tachometer terminal and ground does not flash. Which of the following diagnostic tests should the technician do next?
    A. Replace ICM.
    B. Check battery-voltage feed from the ignition switch to the positive side of the coil.
    C. Check secondary coil windings.
    D. Check primary coil windings.   (C.7 and C.8)

48. A vehicle has poor fuel economy and an excessive odor from the catalytic converter. Technician A says a faulty IAC could be the root cause. Technician B says that the air pump always pumping upstream could be the cause. Who is right?
    A. Technician A only
    B. Technician B only
    C. Both A and B
    D. Neither A nor B   (E.9 and E.10)

49. Refer to the composite vehicle. Technician A says the warm-up cycle is complete when the engine temperature reaches a minimum of 160°F and increase at least 40°F. Technician B says before turning on the MIL, most OBD II failures require two consecutive trips to be completed. Who is right?
    A. Technician A only
    B. Technician B only
    C. Both A and B
    D. Neither A nor B   (B.2 and B.3)

50. When a secondary air-injection system is inspected, many of the hoses are found to be burned. How does this affect the engine's emission output?
    A. Lower-than-normal CO and higher-than-normal HC emissions
    B. Lower-than-normal HC and higher-than-normal CO emissions
    C. Lower-than-normal HC and CO emissions
    D. Higher-than-normal HC and CO emissions   (E.9, F.1 and F.6)

51. During an onboard diagnostic I/M test a vehicle will not communicate with the scan tool. Technician A says the key must be in the ON position for the two computers to communicate. Technician B says the terminals at the DLC could spread preventing communication of the two computers. Who is correct?
    A. A only
    B. B only
    C. Both A and B
    D. Neither A nor B   (F.14)

52. Two technicians are the cause of repeated rotor button failure. Technician A says it could be the result of high secondary resistance. Technician B says the condition of the secondary system should have been checked when the rotor was first replaced. Who is right?
    A. Technician A only
    B. Technician B only
    C. Both A and B
    D. Neither A nor B

    (C.14)

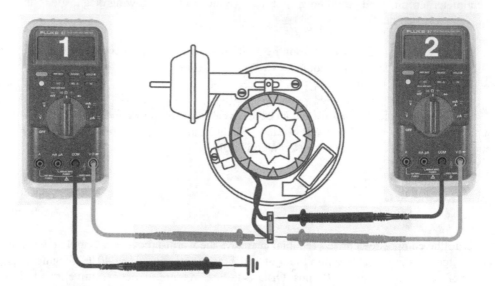

53. Use the figure to determine which of the following pickup coil ohmmeter measurements would be considered normal.
    A. Meter 1, 0 ohms
    B. Meter 2, 0 ohms
    C. Meter 1, infinite ohms
    D. Meter 2, infinite ohms

    (C.8)

54. Which of the following pieces of service information is LEAST necessary for diagnosing an MFI system problem?
    A. Vehicle specifications
    B. Vehicle recall letters
    C. Service manuals
    D. Service bulletins

    (D.2)

55. Technician A says that computerized engine controls can compensate for minor engine misadjustments. Technician B says a scan tool can be used to perform a compression test. Who is right?
    A. Technician A only
    B. Technician B only
    C. Both A and B
    D. Neither A nor B

    (B.2 and B.3)

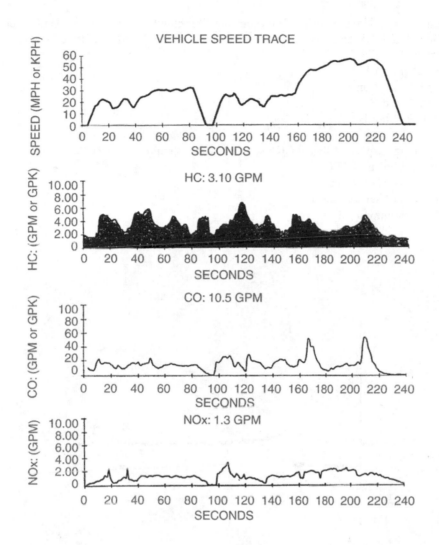

CUTPOINTS		
HC	CO	NOx
Measured in GPM	Measured in GPM	Measured in GPM
0.8	15	2.0

56. Refer to the emission traces in the figure. Which of the following should the technician do next
    to properly diagnose this vehicle?
    A. Monitor sensor input data for invalid inputs
    B. Perform compression test
    C. Perform injector balance test
    D. Check for higher-than-normal fuel pressure   (A.5, A.9 and F.9)

57. Technician A says that an inoperative torque converter clutch could result in a stalling condition. Technician B says that an inoperative torque converter clutch could increase emissions output. Who is right?
    A. Technician A only
    B. Technician B only
    C. Both A and B
    D. Neither A nor B   (A.10)

58. When checking suspect CMP and CKP sensors, which should be checked first?
    A. Connections at the PCM
    B. Connections at the sensors
    C. Strength of sensor magnet
    D. Magnetic timing offset   (B.1)

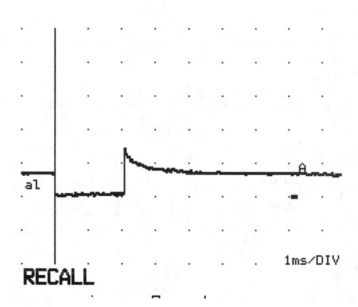

59. Using the injector waveform as a reference, what diagnosis should be made concerning injector operation?
    A. The injector open too long
    B. Low supply voltage
    C. Shorted injector coil
    D. Normal operation   (D.2 and D.11)

60. A vehicle exhibits a rough idle and runs fine at any other RPM. The long-term fuel trim is +25% and the short term fuel trim is +15%. The oxygen sensor is staying a little on the lean bias side. Technician A says to check for a vacuum leak. Technician B says to test the fuel pump for proper operation. Who is correct?
    A. A only
    B. B only
    C. Both A and B
    D. Neither A nor B   (D.4)

61. Technician A says closed loop occurs when the engine is warming up, using its own programming to control air/fuel mixture. Technician B says that closed loop occurs after the engine has reached sufficient temperature and is using O_2 sensor input to determine air/fuel mixture. Who is right?
    A. Technician A only
    B. Technician B only
    C. Both A and B
    D. Neither A nor B   (B.2 and B.3)

62. The recommended service intervals for emission components can be found in:
    A. owners' manuals.
    B. parts location books.
    C. vacuum hose diagrams.
    D. diagnostic procedures.   (E.2)

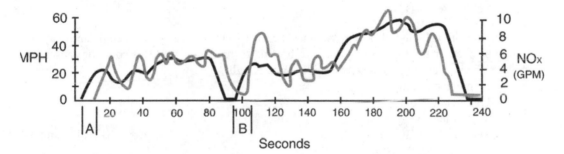

63. The figure shows an NOx emissions trace superimposed over a drive trace. Referring to sections A and B, which of the following terms best describes these trace differences?
    A. Cutpoints
    B. Transients
    C. Delays
    D. Deceleration enrichment   (F.11)

64. O_2 sensor voltage is continually lower than specified. Technician A says this could cause higher-than-normal injector pulse width. Technician B says this condition could cause lower-than-normal block-learn numbers (long-term fuel trim). Who is right?
    A. Technician A only
    B. Technician B only
    C. Both A and B
    D. Neither A nor B   (D.6)

65. A vehicle has just been repaired after failing the vehicle emissions test. Technician A says the vehicle should be driven and then retested before being returned to the customer. Technician B says any stored diagnostic trouble codes should be erased before returning the vehicle to the customer. Who is correct?
    A. A only
    B. B only
    C. Both A and B
    D. Neither A nor B   (E.15)

66. Fuel system diagnostics are being discussed. Technician A says that current ramping of the fuel pump is an effective method of early fuel pump failure detection. Technician B says that current ramping of the fuel pressure regulator is an effective way of diagnosing a bad fuel pressure regulator. Who is correct?
    A. A only
    B. B only
    C. Both A and B
    D. Neither A nor B   (D.7)

67. Air induction systems are being discussed. Technician A says that the manifold turning valve is mainly for wide-open throttle operation. Technician B says tune port air induction systems increase volumetric efficiency. Who is correct?
    A. A only
    B. B only
    C. Both A and B
    D. Neither A nor B   (D.8)

68. The composite vehicle fails an emission test because of excessive NOx. Technician A says this could be because the EGR passages are restricted. Technician B says this could be caused by a hole in the vacuum line to the EGR valve. Who is right?
    A. Technician A only
    B. Technician B only
    C. Both A and B
    D. Neither A nor B   (E.7, E.8 and F.11)

Scan Tool Parameter	Units Displayed	Measured Data Value
Desired EGR	Percent	60%
Actual EGR	Percent	40%

69. Which of the following conditions best fits the data in the table?
    A. Low combustion chamber temperatures
    B. Low NO_x emissions
    C. High NO_x emissions
    D. Cylinder misfire   (E.5, E.6 and F.11)

70. Excessive amounts of alcohol in the fuel system may cause all of the following problems **EXCEPT:**
    A. fuel system corrosion.
    B. engine no-start condition.
    C. rich air/fuel mixture.
    D. fuel filter plugging.   (D.9)

71. A vehicle exhibits an extremely rough idle. During the technician's visual inspection the technician taps the exhaust gas recirculation valve with a small hammer and the vehicle begins to idle smoothly. Technician A says the valve should be replaced. Technician B says the valve just needs to be cleaned. Who is right?
    A. Technician A only
    B. Technician B only
    C. Both A and B
    D. Neither A nor B   (E.11)

72. A leaking cold-start injector can cause all of the following problems **EXCEPT:**
    A. engine stalling.
    B. rough idle.
    C. hard starting.
    D. detonation.   (D.9 and D.10)

73. A customer with a TBI four-cylinder engine complains that on rainy, cool days his vehicle will stall at stops and be difficult to restart. When the vehicle is parked for ten minutes or so, the symptoms will disappear for a short time and then return. Which of the following is the MOST likely cause of the problem?
    A. Fuel pressure too high
    B. Weak ignition coil
    C. Malfunctioning heated air-intake system
    D. Over-advanced ignition timing   (D.9)

74. When troubleshooting an SFI system with low fuel pressure, Technician A says the problem could be a pinched or restricted fuel return line to the fuel tank. Technician B says that low supply voltage to the in-tank electric fuel pump could cause the condition. Who is right?
    A. Technician A only
    B. Technician B only
    C. Both A and B
    D. Neither A nor B   (D.10)

75. A vehicle has just failed a I/M 240 test because of high emissions due to a failed oxygen sensor. Technician A says oxygen sensor activity can be observed with a lab scope. Technician B says remove the oxygen sensor and Ohm out the signal wire and ground with an ohmmeter. Who is right?
    A. Technician A only
    B. Technician B only
    C. Both A and B
    D. Neither A nor B   (F.5)

76. MAP sensor calibration is being discussed. Technician A says some manufacturers require heating the MAP sensor for accurate testing. Technician B says to install a vacuum gauge to the engine and compare the scan tool data to the vacuum gauge. Who is correct?
    A. A only
    B. B only
    C. Both A and B
    D. Neither A nor B   (B.18)

77. Refer to the composite vehicle schematic. There is a higher-than-normal idle and the A/C compressor clutch won't engage in the composite vehicle. Technician A says PCM terminal 63 could be shorted to power. Technician B says the switch in circuit for PCM terminal 65 is closed and is the problem. Who is right?
    A. Technician A only
    B. Technician B only
    C. Both A and B
    D. Neither A nor B   (B.2, B.4, B.13, B.17 and B.20)

78. High fuel pressure in an SFI system can result from all of the following **EXCEPT:**
    A. a plugged or restricted fuel return line.
    B. a sticking pressure regulator.
    C. excessive voltage drop on the fuel pump ground.
    D. incorrectly routed or leaking vacuum hoses.   (D.10)

79. Which of the following component could MOST likely restrict airflow enough to reduce power from the affected cylinders during cruise or acceleration?
    A. Worn piston rings
    B. Worn camshaft lobes
    C. Worn cylinder walls
    D. Worn valve guides   (A.9)

Cylinder	1	2	3	4	5	6
High Reading	225	225	225	225	225	225
Low Reading	100	100	100	90	100	115
Amount of Drop	125	125	125	135	125	110
Injector Condition	OK	OK	OK	?	OK	?

80. Use the chart to interpret injector pressure-drop test data. Technician A says the injector in cylinder 4 is faulty and running too lean. Technician B says the injector in cylinder 6 is faulty and running too rich. Who is right?
    A. Technician A only
    B. Technician B only
    C. Both A and B
    D. Neither A nor B   (D.11)

81. Which of the following would NOT be included in a preliminary emission system inspection?
    A. OCV valve and connections
    B. Proper vacuum hose connections
    C. Catalytic converter and exhaust installation
    D. Catalytic converter inlet/outlet temperature   (E.1)

82. Which of the following diagnostic tools should be used to check for secondary output of an ignition coil?
    A. Voltmeter
    B. Ohmmeter
    C. Ignition module tester
    D. Test spark plug   (C.7 and C.8)

83. Which of the following problems would LEAST likely cause a continuing misfire condition in a distributor-ignition system?
    A. Low engine compression
    B. Combustion chamber carbon deposits
    C. Intake manifold vacuum leaks
    D. Worn distributor bushings   (C.8)

84. Which of the following symptoms could be associated with an EGR valve stuck closed?
    A. Hard starting
    B. Rough idle
    C. Engine stalling
    D. Detonation   (E.6)

85. While performing a tune up on a vehicle, the technician finds an oil-soaked air filter. Which of the following is the MOST likely cause of this condition?
    A. Excessively rich air/fuel mixture
    B. EGR valve stuck closed
    C. Restricted catalytic converter
    D. PCV valve stuck closed.   (E.7 and E.8)

86. Two technicians are discussing CAN communication. Technician A says that an open in the CAN bus will cause a no start. Technician B says CAN is the communication network used by all OBD II systems. Who is right?
    A. Technician A only
    B. Technician B only
    C. Both A and B
    D. Neither A nor B   (B.4)

87. The I/M240 test interprets average emission levels over a predetermined period of time. What scale of measurement is used to display the emissions average?
    A. ppm
    B. gpm
    C. Ratio
    D. Percentage   (F.3)

88. Which of the following is LEAST useful when inspecting for tampering?
    A. Scan tool
    B. Parts location book
    C. Wiring diagram
    D. Service bulletins   (A.1)

89. Which of the following memory devices loses its memory when the ignition is switched off?
    A. Volatile RAM
    B. Nonvolatile RAM
    C. ROM
    D. PROM   (B.2)

SCAN TOOL DATA			
Engine Coolant Temperature (ECT) Sensor 205°F/ 96°C/ 0.52 volts	Intake Air Temperature (IAT) Sensor 100°F/ 38°C/ 2.4 volts	Mass Airflow Sensor (MAF) 8 gm/sec/1.50 volts	Throttle Actuator Control Motor (TAC) 35 percent
Throttle Position Sensor 1 (TP1) 0 percent / 5.00 volts	Throttle Position Sensor 2 (TP2) 35 percent / 1.60 volts	Accelerator Pedal Position Sensor 1 (APP1) 35 percent / 1.40 volts	Accelerator Pedal Position Sensor 2 (APP2) 35 percent / 2.4 volts
Crankshaft Position Sensor (CPS) 1500 rpm	Heated Oxygen Sensor Bank 1 (HO2S 1/1) .01-.90 volts	Heated Oxygen Sensor Bank 2 (HO2S 2/1) 0.1-.90 volts	Heated Oxygen Sensor Post-Cat (HO2S 1/2) 0.4 volts
Battery Voltage (B+) 14.1 volts	EVAP Canister Purge Solenoid 0 percent	EVAP Canister Vent Solenoid OFF	Fuel Pump Relay (FP) ON
Vehicle Speed Sensor (VSS) 40 MPH	Open/Closed Loop OPEN	Malfunction Indicator Lamp (MIL) ON	Ignition Timing Advance 12 BTDC

Measured Ignition Timing °BTDC   Base Timing 10°   Actual Timing 11°

90. The composite vehicle has experienced a lack of power complaint. The vehicle also has a confirmed TPS DTC. Using the data shown, what is the MOST likely cause of these failures?
    A. Bad TPS 1
    B. Bad TPS 2
    C. Bad APP1
    D. Bad APP2   (B.10)

91. During a road test, the composite vehicle cuts out and loses power when engine speed rises above 6,000 rpm. Technician A says a restricted fuel line or restricted fuel filter may be the cause. Technician B says a bad ignition module may be the cause. Who is right?
    A. Technician A only
    B. Technician B only
    C. Both A and B
    D. Neither A nor B   (B.2, B.3 and D.10)

92. A vehicle demonstrates a part-throttle surge. All normal diagnostic procedures have failed to locate the cause. Technician A says the powertrain control module (PCM) should be replaced. Technician B says to consult the manufacturer's service bulletins for calibration change information. Who is right?
    A. Technician A only
    B. Technician B only
    C. Both A and B
    D. Neither A nor B   (A.2 and B.2)

93. A vehicle with electronic engine controls has a deceleration stalling condition. All initial checks show no cause for this condition, and the technician has verified the complaint. Which of the following should the technician do next?
    A. Consult the manufacturer's service bulletin updates.
    B. Replace the powertrain control module (PCM).
    C. Recommend high-octane fuel to the customer.
    D. Adjust the idle speed.   (B.2 and B.3)

6 Additional Test Questions for Practice

Additional Test Questions

Please note the letter and number in parentheses following each question. They match the task in Section 4 that discusses the relevant subject matter. You may want to refer to the overview using the cross-referencing key to help with questions posing problems for you.

1. Before diagnosing an SFI system, a preliminary diagnostic procedure must be performed. This procedure includes all of the following **EXCEPT:**
   A. an emissions test.
   B. a fuel pressure test.
   C. a check for vacuum leaks.
   D. a tamper inspection. (D.1, D.9 and D.10)

2. An engine with computerized fuel injection has a burned exhaust valve, resulting in low-cylinder compression. Technician A says a scan tool will show this as a continuous lean exhaust condition. Technician B says a scan tool may show higher-than-normal long-term fuel trim command. Who is right?
   A. Technician A only
   B. Technician B only
   C. Both A and B
   D. Neither A nor B (A.9)

3. When testing a suspect knock sensor, the technician should first:
   A. check the knock sensor for an open condition.
   B. check the timing while running at 2,000 rpm for a tapping beside sensor.
   C. test for a shorted condition.
   D. backprobe the knock sensor connector at the control unit for voltage. (B.9 and B.10)

4. A vehicle fails a state emission test for excess CO at 2,500 rpm. All of the following could cause this EXCEPT:
   A. a saturated charcoal canister.
   B. air injection stuck in the upstream mode.
   C. a vacuum leak.
   D. fuel-diluted engine oil. (F.3 and F.8)

5. If a vehicle with drivability problems runs better in "limp-home" mode than it does normally, which of the following is the MOST likely cause?
   A. PCM
   B. MAF sensor
   C. O_2 sensor
   D. Catalytic converter (B.1 and B.8)

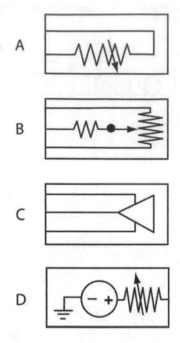

6. In the figure, which symbol represents the most common type of ECT sensor?
   A. A
   B. B
   C. C
   D. D   (B.2)

7. Which of the following is not currently a scan tool data item?
   A. Barometric pressure
   B. O_2 sensor voltage
   C. Fuel octane
   D. Miles per hour   (B.2)

8. What type of signal is produced by a typical piezoelectric knock sensor (KS)?
   A. A change in resistance
   B. A change in capacitance
   C. An oscillating analog voltage signal
   D. A switching digital voltage signal   (C.5, C.9 and C.10)

9. Most emissions analyzers measure gases and display them as either parts per million (PPM)
   or as a percentage of total exhaust gas volume. Which of the following emission gases are
   usually measured in PPM?
   A. CO_2
   B. O_2
   C. CO
   D. HC   (F.3)

10. With the radiator cap removed and the engine running, a four-gas analyzer can be used to:
    A. test for a blown head gasket.
    B. check for a burned exhaust valve.
    C. check for a burned intake valve.
    D. check engine vacuum.   (A.9)

11. Running a DI-equipped vehicle with a spark plug wire removed could result in damage to all of the following components **EXCEPT** the:
    A. catalytic converter.
    B. ignition coil.
    C. ignition module.
    D. ignition switch.   (C.2, C.8 and C.12)

12. A vehicle has a miss at idle. Gas analysis shows elevated HC emissions, CO_2 at approximately 11%, O_2 at approximately 6%, and CO at less than 1%. Which of the following is the MOST likely cause of this condition?
    A. A leaking injector
    B. A restricted air filter
    C. Excessive fuel pressure
    D. A vacuum leak   (F.3 and F.4)

13. When interpreting scan tool serial data, desired rpm is defined as:
    A. measured engine speed.
    B. cranking rpm.
    C. engine speed commanded by the PCM.
    D. throttle angle.   (B.2)

14. Technician A says that a gas analyzer can be used to find head gasket or combustion chamber leaks by sampling the area over an open radiator of a running engine. Technician B says an EVAP system leak can be found by using the gas analyzer and watching for an HC reading increase when the probe is near the leak. Who is right?
    A. Technician A only
    B. Technician B only
    C. Both A and B
    D. Neither A nor B   (A.7 and A.8)

15. Technician A says CO readings are a good indicator of a rich air/fuel ratio. Technician B says CO readings also are a good indicator of a lean air/fuel ratio. Who is right?
    A. Technician A only
    B. Technician B only
    C. Both A and B
    D. Neither A nor B   (F.4)

16. Technician A says a rough-running engine can be caused by leaking valves and worn piston rings. Technician B says a leaking head gasket can cause a rough running condition. Who is right?
    A. Technician A only
    B. Technician B only
    C. Both A and B
    D. Neither A nor B   (A.9)

SCAN TOOL DATA			
Engine Coolant Temperature (ECT) Sensor 230°F/ 11°C/ 0.35 volts	Intake Air Temperature (IAT) Sensor 66°F/ 19°C/ 3.0 volts	Mass Airflow Sensor (MAF) 0 gm/sec/.20 volts	Throttle Actuator Control Motor (TAC) 0 percent
Throttle Position Sensor 1 (TP1) 0 percent / 4.50 volts	Throttle Position Sensor 2 (TP2) 0 percent / 0.50 volts	Accelerator Pedal Position Sensor 1 (APP1) 0 percent / 0.50 volts	Accelerator Pedal Position Sensor 2 (APP2) 0 percent / 1.50 volts
Crankshaft Position Sensor (CPS) 0 rpm	Heated Oxygen Sensor Bank 1 (HO2S 1/1) .00 volts	Heated Oxygen Sensor Bank 2 (HO2S 2/1) 0.0 volts	Heated Oxygen Sensor Post-Cat (HO2S 1/2) 0.0 volts
Battery Voltage (B+) 12.3 volts	EVAP Canister Purge Solenoid 0 percent	EVAP Canister Vent Solenoid OFF	Fuel Pump Relay (FP) ON
Vehicle Speed Sensor (VSS) 0 MPH	Open/Closed Loop OPEN	Malfunction Indicator Lamp (MIL) OFF	Ignition Timing Advance 0 BTDC

Measured Ignition Timing °BTDC   Base Timing 10°   Actual Timing 12°

17. The composite vehicle is difficult to start when cold, runs poorly, and backfires through the intake when trying to accelerate. Once the vehicle is warm, it performs normally. The data in the table was obtained with the key-on and engine-off (KOEO) after the vehicle had been outside overnight. How should the technician proceed?
    A. Replace the oxygen sensor.
    B. Check the ECT and its circuit.
    C. Check the IAT sensor and circuit.
    D. Check PCM power and ground.   (B.2, B.17 and B.20)

18. A fuel-injected vehicle misfires and cuts out at steady cruise with very light throttle application. The vehicle operates normally in all other conditions. Which of the following is the MOST likely cause of this problem?
    A. Worn spark plugs
    B. Dirty fuel injectors
    C. Faulty knock sensor
    D. Defective TPS   (B.10 and B.17)

19. All of the following can lead to excessive HC emissions EXCEPT:
    A. ignition system misfire.
    B. low cylinder compression.
    C. excessively lean air/fuel mixture.
    D. leaking exhaust system.   (A.10 and F.7)

20. A vehicle passes an I/M emission test for HC and CO but fails for high NOx emissions. All of the following could be the cause EXCEPT:
    A. excessive carbon buildup on the piston heads.
    B. a plugged EGR port.
    C. hot spots in the combustion chamber caused by cooling system deposits.
    D. a rich air/fuel mixture.   (F.4, F.6 and F.11)

21. Battery voltage is found to be 12.6 volts; but when checking supply voltage at the PCM, only 12.0 volts is measured. Technician A says to check the PCM ground connections. Technician B says to check for high resistance in the PCM supply circuit. Who is right?
    A. Technician A only
    B. Technician B only
    C. Both A and B
    D. Neither A nor B   (B.13)

22. The intake air temperature (IAT) scan tool serial data indicates:
    A. temperature required by the PCM.
    B. temperature required for closed-loop operation.
    C. temperature signal as received by the PCM.
    D. temperature signal as sent by the IAT sensor.   (B.2, B.3 and B.4)

23. A technician is diagnosing a vehicle that is hard to restart when the engine is warm. Typically, the owner shuts off the vehicle, returns after about 15 minutes, and tries to start the vehicle. The engine cranks normally but won't start. If the owner waits about two hours, the vehicle starts normally. Which of the following measurements is LEAST likely to lead to a correct diagnosis of the problem?
    A. Fuel pressure and volume
    B. Battery load capacity
    C. Ignition coil resistance
    D. Ignition pickup waveform   (C.4, C.8 and C.10)

24. If the composite vehicle has a coolant temperature of 100°C, what voltage would you expect to see from the ECT?
    A. 3.59 volts
    B. 2.27 volts
    C. 0.46 volts
    D. 0.84 volts   (B.2 and B.17)

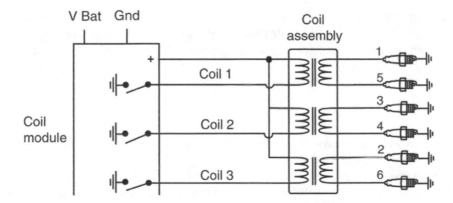

25. Refer to the schematic in the figure. Technician A says that the primary windings of coils 1, 2, and 3 are wired in parallel; and spark plugs 1 and 5 are wired in series. Technician B says coils 1, 2, and 3 are wired in parallel and spark plugs 1 and 5 are fired together as a pair. Who is right?
    A. Technician A only
    B. Technician B only
    C. Both A and B
    D. Neither A nor B   (C.2 and C.3)

26. Computer output driver circuits may control current to all of the following devices **EXCEPT**:
    A. relays.
    B. solenoids.
    C. sensors.
    D. displays.   (B.2 and B.3)

27. In an OBD II system, the malfunction indicator lamp (MIL) lights when emissions exceed:
    A. 2.0 times the allowable standard.
    B. 2.0% over the allowable standard.
    C. 1.5 time the allowable standard.
    D. 1.5% over the allowable standard.   (B.2 and B.3)

28. What action should the PCM take when it receives a high-voltage signal from the oxygen sensor?
    A. Extend the IAC pintle
    B. Increase fuel delivery
    C. Decrease fuel delivery
    D. Energize the purge solenoid   (B.2 and B.3)

29. Higher than normal HC emissions can result from all of the following **EXCEPT**:
    A. EGR valve failure
    B. cylinder misfire
    C. incomplete combustion
    D. an excessively lean air/fuel mixture   (A.5, E.8 and F.6)

30. A TBI-injected vehicle requires extended cranking time to start when parked for long periods
    of time, such as overnight. Which of the following problems would LEAST likely cause this
    condition?
    A. Leaking fuel pump check valve
    B. Coolant sensor out of calibration
    C. Bad PCM
    D. PCM not receiving a cranking input   (B.17, B.20, D.9 and D.10)

31. Technician A says that most TPS and MAP sensors operate on a 5-volt reference voltage supply
    from the PCM. Technician B says that analog TPS and MAP sensors deliver voltage signals that
    range from less than 1 volt at idle to almost reference voltage at wide-open throttle. Who is right?
    A. Technician A only
    B. Technician B only
    C. Both A and B
    D. Neither A nor B   (B.2, B.3 and B.17)

32. With a scan tool connected to a vehicle, the technician can do all of the following **EXCEPT**:
    A. change resistance values for sensors.
    B. read trouble codes.
    C. compare key-on, engine-off (KOEO) data to key-on, engine-running (KOER) data.
    D. watch sensor outputs through their ranges.   (B.2 and B.3)

33. A high O_2 sensor voltage is indicating a rich exhaust condition, and the injector pulse width is
    lower than normal. All of the following factors could cause this **EXCEPT**:
    A. a faulty ECM.
    B. leaking injectors.
    C. excessive fuel pressure.
    D. the EVAP system purging at idle.   (B.2 and B.3)

SCAN TOOL DATA			
Engine Coolant Temperature (ECT) Sensor 212°F/ 100°C/ 0.46 volts	Intake Air Temperature (IAT) Sensor 102°F/ 39°C/ 2.24 volts	Mass Airflow Sensor (MAF) 5 gm/sec/1.1 volts	Throttle Actuator Control Motor (TAC) 0 percent
Throttle Position Sensor 1 (TP1) 0 percent / 4.50 volts	Throttle Position Sensor 2 (TP2) 0 percent / 0.50 volts	Accelerator Pedal Position Sensor 1 (APP1) 0 percent / 0.50 volts	Accelerator Pedal Position Sensor 2 (APP2) 0 percent / 1.50 volts
Crankshaft Position Sensor (CPS) 1500 rpm	Heated Oxygen Sensor Bank 1 (HO2S 1/1) .02-0.5 volts	Heated Oxygen Sensor Bank 2 (HO2S 2/1) 0.2-0.5 volts	Heated Oxygen Sensor Post-Cat (HO2S 1/2) 0.1-0.3 volts
Battery Voltage (B+) 14.2 volts	EVAP Canister Purge Solenoid 0 percent	EVAP Canister Vent Solenoid OFF	Fuel Pump Relay (FP) ON
Vehicle Speed Sensor (VSS) 0 mph	Open/Closed Loop CLOSED	Malfunction Indicator Lamp (MIL) OFF	Ignition Timing Advance 10 BTDC
Transmission Range Switch PARK	Trans Pressure Control Solenoid 80%	Transmission Shift Solenoid 1 ON	Transmission Shift Solenoid 2 OFF
Fuel Level Sensor 3.5 volts	Fuel Tank Pressure Sensor 2.0 volts	Transmission Fluid Temperature Sensor 0.6 volts	Transmission Turbine Shaft Speed Sensor 0 mph
Measured Ignition Timing °BTDC	Base Timing 10°	Actual Timing 20°	

34. Refer to the scan data. The composite vehicle has no other drivability complaints other than what is evidenced by the rpm and TAC readings above. What is the MOST likely cause of this condition?
    A. The timing belt has jumped timing.
    B. Ignition timing is too retarded.
    C. Fuel pressure is too high.
    D. There is a vacuum leak at the intake manifold.   (A.9, B.2, B.3 and B.4)

35. The composite vehicle will not communicate with a scan tool. Technician A says if terminal 16 at the data link connector becomes shorted, fuse #74 will fail and cause this condition. Technician B says the PCM will MOST likely have a DTC and a snapshot or freeze frame stored in memory to help in diagnosis. Who is right?
    A. Technician A only
    B. Technician B only
    C. Both A and B
    D. Neither A nor B   (B.2, B.3, B.4 and B.13)

36. Repeated computer problems can result from excessive radiofrequency interference (RFI). Which of the following tools would be most useful in diagnosing this problem?
    A. Diode tester
    B. Ohmmeter
    C. Voltmeter
    D. Oscilloscope   (B.1, B.3 and B.13)

37. Vehicle-specific emission vacuum hose routing information can be found:
    A. in the vehicle owner's guide
    B. in the trunk area
    C. in the vehicle's audio guide
    D. under the hood on the emission label   (A.2 and E.2)

38. A vehicle with OBD II has a no-start condition. Preliminary checks show that the fuel injectors
    are not working. The DTCs are obtained, and one of them is P0123-TPS circuit high. Technician
    A says voltage on the TPS sensor over approximately 3.0 volts may cause the vehicle to go into
    a clear-flood mode, disabling the fuel injectors. Technician B says the fuel pump has failed
    and will have to be replaced. Who is right?
    A. Technician A only
    B. Technician B only
    C. Both A and B
    D. Neither A nor B   (B.2, B.4, B.10 and D.3)

39. Technician A says that shorted output actuators can cause repeated computer failures. Technician
    B says that faulty voltage-suppression diodes can cause repeated computer failures. Who is right?
    A. Technician A only
    B. Technician B only
    C. Both A and B
    D. Neither A nor B   (B.2, B.3, B.10 and B.20)

40. While discussing catalytic converter operation, Technician A says if HC levels entering the
    converter remain too high for too long, damage may result. Technician B says the converter uses
    oxygen in the process of converting HC and CO to CO_2 and H_2O. Who is right?
    A. Technician A only
    B. Technician B only
    C. Both A and B
    D. Neither A nor B   (E.2 and E.3)

41. The "stoichiometric" mixture refers to:
    A. the air/fuel mixture of 14.7 parts air, 1 part fuel.
    B. the air/fuel mixture that yields the best power.
    C. the air/fuel mixture that yields the best economy.
    D. the air/fuel mixture that yields the lowest emissions.   (E.2 and E.3)

42. Which of the following conditions would LEAST likely cause a PO300 random misfire code on
    an OBD II vehicle?
    A. Excessive Engine oil pressure
    B. Broken rocker arm shaft pedestal
    C. Leaking fuel-pressure regulator
    D. Low alternator output   (A.2, A.6, A.9 and D.10)

43. A vehicle with sequential fuel injection had the throttle body clean and serviced less than 3
    months ago for stalling and is in need of another throttle body cleaning for stalling. Technician
    A says the vehicle may have excessive blow-by. Technician B says the stalling can't be because
    of a dirty throttle body. Who is right?
    A. Technician A only
    B. Technician B only
    C. Both A and B
    D. Neither A nor B   (D.14)

44. Scan tool serial data displays a coolant temperature of 55°F with the engine at normal operating temperature. Technician A says to check the ECT circuit wiring and connections using a DVOM, including power, ground, and voltage-drop tests. Technician B says to pull the ECT sensor and check it in hot water with an ohmmeter. Who is right?
    A. Technician A only
    B. Technician B only
    C. Both A and B
    D. Neither A nor B   (B.4 and B.13)

45. Which of the following conditions is LEAST likely to cause a loss of power while accelerating?
    A. A vacuum leak
    B. A clogged fuel filter
    C. An overtorqued knock sensor
    D. Poor fuel quality   (C.11 and D.9)

46. An OBD II-compliant vehicle has an illuminated MIL with a PO101 mass airflow sensor performance code stored. Technician A says a faulty mass airflow sensor could be the cause. Technician B says a mechanical problem such as a restricted exhaust system could be the cause. Who is right?
    A. Technician A only
    B. Technician B only
    C. Both A and B
    D. Neither A nor B   (A.9, B.2, B.3, B.4 and B.20)

47. A vehicle fails an I/M emission test for HC at idle. The four-gas analyzer readings at idle and at 2,000 rpm are as follows:

	Idle	2,000 rpm
HC (ppm)	700	20
CO (%)	0.45	0.20
CO_2 (%)	12.00	14.60
O_2 (%)	2.50	0.90

Additionally, the car idles roughly but smoothes out as engine speed increases. The MOST likely cause is:
    A. a shorted fuel-injector driver.
    B. worn intake cam lobe.
    C. an intake vacuum leak.
    D. a fouled spark plug.   (E.8, F.6 and F.7)

48. A vehicle failed a loaded-mode CO emissions test. The O_2 sensor voltage is at 850 mV. All items below may cause this **EXCEPT:**
    A. engine-coolant temperature sensor.
    B. oxygen sensor (O_2).
    C. throttle position sensor (TPS).
    D. manifold absolute pressure sensor (MAP).   (B.10 and F.10)

49. Technician A says that a fuel-injection block learn reading (long-term fuel trim) that is low at idle and at 3,000 rpm indicates an overall rich-running engine. Technician B says that a block learn reading (long-term fuel trim) that is high at idle and normal at 3,000 rpm can indicate a vacuum leak. Who is right?
    A. Technician A only
    B. Technician B only
    C. Both A and B
    D. Neither A nor B   (B.10, D.6 and D.9)

50. Technician A says that maximum secondary coil voltage is necessary to compensate for high cylinder pressures at wide-open throttle (WOT). Technician B says secondary reserve coil voltage is required to fire a stoichiometric (14.7:1) air/fuel ratio. Who is right?
    A. Technician A only
    B. Technician B only
    C. Both A and B
    D. Neither A nor B   (C.3 and C.8)

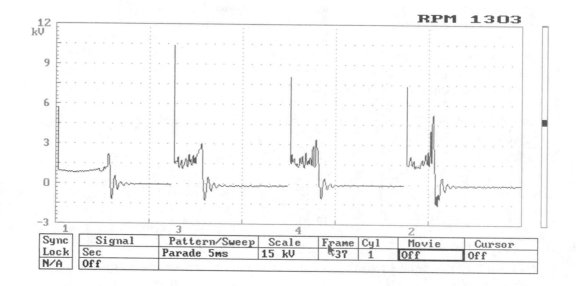

51. All of the following conditions could cause the problem seen in the ignition waveform shown **EXCEPT:**
    A. wide spark plug gap.
    B. fouled spark plug.
    C. low cylinder compression.
    D. bridged spark plug gap.   (C.9, C.10 and C.12)

52. An SFI-equipped vehicle fails an enhanced I/M emissions test for high CO. Technician A says that a faulty fuel-pressure regulator could be the cause. Technician B says that repairing the problem causing high CO emissions could result in a retest failure for NO_X. Who is right?
    A. Technician A only
    B. Technician B only
    C. Both A and B
    D. Neither A nor B   (D.9, F.2 and F.4)

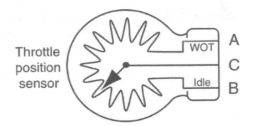

53. Using the figure for reference, Technician A says the computer supplies 5 volts to the sensor at terminal A and grounds terminal B. Technician B says with the sensor wiper in the position shown (idle), signal voltage at terminal C should be low and increase as the wiper rotates clockwise. Who is right?
    A. Technician A only
    B. Technician B only
    C. Both A and B
    D. Neither A nor B   (B.10)

54. When using supply voltage to test computer system solenoid windings that incorporate voltage suppression diodes: Technician A says normal system polarity must be observed during testing. Technician B says the solenoid will energize regardless of polarity without damage. Who is right?
    A. Technician A only
    B. Technician B only
    C. Both A and B
    D. Neither A nor B   (B.10)

55. A vehicle with an AIR system is found to have the air pipe feeding the catalytic converter broken off at the converter. Technician A says this problem will increase HC, CO, and NO_x emissions. Technician B says this problem only affects emissions during warm-up operation. Who is right?
    A. Technician A only
    B. Technician B only
    C. Both A and B
    D. Neither A nor B   (E.2, E.3 and E.9)

56. Which of the following is used to cool combustion chamber temperatures below 2,500°F?
    A. HC
    B. NOx
    C. Exhaust gas
    D. Air   (E.7, E.9 and F.11)

57. Two technicians are discussing diagnosing a no-code drivability problem. Technician A says to check all computer ground connections with a DMM first. Technician B says the service history of the vehicle should be reviewed for any recent repairs and the problem verified before performing pinpoint tests. Who is right?
    A. Technician A only
    B. Technician B only
    C. Both A and B
    D. Neither A nor B   (B.4 and B.13)

58. A vehicle has a power loss and poor fuel mileage. Technician A says that late ignition timing could be the cause. Technician B says excessive centrifugal or vacuum advance could be the cause. Who is right?
    A. Technician A only
    B. Technician B only
    C. Both A and B
    D. Neither A nor B   (C.7 and C.11)

59. On a vehicle with a computer-controlled carburetor, the throttle position sensor is stuck in the W.O.T. position. Technician A says the vehicle may operate with a fixed rich command to the mixture control solenoid. Technician B says the vehicle may fail a loaded mode emissions test. Who is right?
    A. Technician A only
    B. Technician B only
    C. Both A and B
    D. Neither A nor B   (D.5, D.6, F.9 and F.10)

SCAN TOOL DATA

Engine Coolant Temperature (ECT) Sensor 210°F/ 99°C/ 0.51 volts	Intake Air Temperature (IAT) Sensor 95°F/ 35°C/ 2.4 volts	Mass Airflow Sensor (MAF) 15 gm/sec/2.0 volts	Throttle Actuator Control Motor (TAC) 50 percent
Throttle Position Sensor 1 (TP1) 50 percent / 2.50 volts	Throttle Position Sensor 2 (TP2) 50 percent / 2.50 volts	Accelerator Pedal Position Sensor 1 (APP1) 50 percent / 2.00 volts	Accelerator Pedal Position Sensor 2 (APP2) 50 percent / 3.00 volts
Crankshaft Position Sensor (CPS) 1700 rpm	Heated Oxygen Sensor Bank 1 (HO2S 1/1) .02-0.8 volts	Heated Oxygen Sensor Bank 2 (HO2S 2/1) 0.2-0.7 volts	Heated Oxygen Sensor Post-Cat (HO2S 1/2) 0.4 volts
Battery Voltage (B+) 14.3 volts	EVAP Canister Purge Solenoid 30 percent	Torque Converter Clutch (TCC) Solenoid ON-OFF-ON-OFF	Fuel Pump Relay (FP) ON
Vehicle Speed Sensor (VSS) 55 MPH	Open/Closed Loop CLOSED	Malfunction Indicator Lamp (MIL) OFF	Ignition Timing Advance 22 BTDC

Measured Ignition Timing °BTDC   Base Timing 10°   Actual Timing 32°

60. The composite vehicle feels as if it has a miss at cruise but the MIL is illuminated. Using the scan tool data, what is the MOST likely cause of this condition?
    A. A faulty oxygen sensor
    B. Low fuel pressure
    C. A faulty fuel injector
    D. Intermittent torque converter clutch solenoid circuit   (B.2 and B.19)

61. An engine idles smoothly but has high HC emissions at idle. CO is within limits. Technician A says that a restricted air filter may be causing the high HC. Technician B says that overly advanced ignition timing at idle can cause high HC emissions. Who is right?
    A. Technician A only
    B. Technician B only
    C. Both A and B
    D. Neither A nor B   (C.11 and F.7)

62. Lower combustion chamber temperatures will result in:
   A. lower CO emissions.
   B. lower CO_2 emissions.
   C. lower NO_x emissions
   D. elevated O_2 emissions.   (E.2 and E.3)

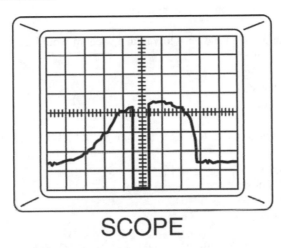

SCOPE

63. A technician is testing a TPS and obtains the trace shown on the DSO in the figure. How should
    the result be interpreted?
   A. This is a normal result and the TPS is in good condition.
   B. The technician did not perform the test properly by not moving the throttle smoothly.
   C. There is an open or high resistance in the TPS, and it must be replaced.
   D. The dip in the trace is the wide-open throttle threshold. The TPS is okay.   (B.17)

64. A vehicle equipped with SFI fails a no-load I/M test with high CO emissions. Checking inputs
    to the PCM reveals a higher-than-normal voltage from the O_2 sensor. Technician A says a
    balance check should be performed on the injectors. Technician B says to check the fuel
    pressure. Who is right?
   A. Technician A only
   B. Technician B only
   C. Both A and B
   D. Neither A nor B   (B.17, D.10, E.10 and F.7)

65. Referring to the composite vehicle schematic information, an open circuit at PCM terminal
    25 may cause all of the following **EXCEPT:**
   A. hard starting and engine stalling when warm.
   B. improper torque converter clutch lock-up.
   C. lean air/fuel mixture.
   D. improper EGR operation.   (B.2, B.10 and B.17)

66. While discussing EVAP purge system operation, Technician A says that the vehicle speed has
    to be below a certain threshold for operation. Technician B says coolant temperature has to be
    below a certain threshold for operation. Who is right?
   A. Technician A only
   B. Technician B only
   C. Both A and B
   D. Neither A nor B   (D.3)

67. Refer to the composite vehicle information. Low fuel volume could be cause by any of the following **EXCEPT:**
    A. restricted vacuum line to the fuel-pressure regulator.
    B. low voltage supply at the fuel pump.
    C. clogged fuel filter.
    D. excessive voltage drop on the fuel pump ground.   (D.10)

68. Engine power loss under load and very low intake vacuum would LEAST likely result from:
    A. A faulty EGR valve.
    B. retarded cam timing.
    C. a restricted exhaust.
    D. a rich air/fuel ratio.   (D.3)

69. In a TBI system, which of the following symptoms would NOT suggest a low-fuel pressure condition?
    A. Engine surging
    B. Black exhaust smoke
    C. Lack of power
    D. Acceleration stumbles   (D.9)

70. A vehicle has failed the I/M test and has a MIL on. The DTC in the PCM is PO442, small leak detected. Technician A says the cap should be pressure tested for leaks. Technician B says use a smoke machine and check for leaks at seal areas such as the filler neck, gas cap, and hose connections at the tank. Who is correct?
    A. A only
    B. B only
    C. Both A and B
    D. Neither A nor B   (F.15)

71. Refer to the Composite Vehicle schematic in Section 4. When diagnosing intermittent injector ground faults, Technician A says the composite vehicle schematic represents a typical multiport group-fired (non-sequential) fuel-injection system. Technician B says that each injector in this system has its own ground circuit. Who is right?
    A. Technician A only
    B. Technician B only
    C. Both A and B
    D. Neither A nor B   (D.2, D.3 and D.10)

72. Which of the following could cause a rough idle, surging, and hesitation in a vehicle with sequential fuel injection (SFI)?
    A. Restricted injectors
    B. Intake manifold vacuum leaks
    C. Intake valve deposits
    D. All of the above   (D.9, D.11 and D.12)

73. During I/M testing, all of the following problems could result in NO_x emissions test failure **EXCEPT:**
    A. a faulty thermotime switch
    B. inoperative EGR valve.
    C. a faulty catalytic converter
    D. a catalytic converter not heated up.   (E.2, E.7, F.2 and F.11)

74. A vehicle with a catalytic converter fails a loaded-mode I/M test with high CO. The CO trace printed out at the conclusion of the test indicates that CO was high only at speeds above 20 mph. Technician A says the PCV valve is stuck closed. Technician B says the charcoal canister is saturated with fuel. Who is right?
    A. Technician A only
    B. Technician B only
    C. Both A and B
    D. Neither A nor B   (E.7, F.10 and F.16)

75. A vehicle experiences repeated oxygen sensor failures. Technician A says that a leaking head gasket may be allowing coolant into the combustion chamber and contaminating the sensor as it passes out in the exhaust. Technician B says that ethylene glycol and its additives can form deposits on the oxygen sensor and prevent it from accurately measuring exhaust oxygen. Who is right?
    A. Technician A only
    B. Technician B only
    C. Both A and B
    D. Neither A nor B   (B.17 and B.21)

76. Referring to the composite vehicle, the technician confirms that the vehicle has poor mileage and a spark knock condition. Technician A says to check the knock sensor using a small hammer and a timing light. Technician B says to adjust ignition timing. Who is right?
    A. Technician A only
    B. Technician B only
    C. Both A and B
    D. Neither A nor B   (C.3)

77. The composite vehicle is shifting late and hard on all upshifts. Technician A says the wire to PCM terminal 76 could be shorted to ground. Technician B says the shifting problem could affect a loaded-mode emissions test. Who is right?
    A. Technician A only
    B. Technician B only
    C. Both A and B
    D. Neither A nor B   (B.4, F.3 and A.10)

78. Carbon buildup on intake valves can result in which of the following air/fuel mixture problems?
    A. Condensation
    B. Absorption
    C. Vaporization
    D. Puddling   (D.9)

79. All of the following are needed to diagnose a general powertrain mechanical failure **EXCEPT:**
    A. relevant technical service bulletins
    B. a scan tool
    C. the vehicle's service manual
    D. a vacuum gauge   (A.2, A.3 and A.11)

| Fused spot deposit | Overheating | Carbon fouled | Pre-ignition |

80. From the examples in the figure shown, which of the following spark plug conditions would LEAST likely be the result of an ignition-related problem?
    A. Overheating
    B. Fused spot deposits
    C. Pre-ignition
    D. Carbon fouling   (C.7 and C.8)

Engine Speed	Idle	2000 RPM
HC (ppm)	25	12
CO (percent)	0.3	0.1
CO_2 (percent)	14	14.5
O_2 (percent)	0.5	0.1

81. Refer to the emission reading chart. What could be said of this vehicle?
    A. The vehicle has good combustion efficiency and catalytic converter action.
    B. The vehicle has a vacuum leak.
    C. The vehicle has a leaking exhaust system.
    D. The catalytic converter is failing.   (E.3 and F.3)

82. A vehicle is being checked for monitor readiness status. Technician A says the scan tool is used for this function. Technician B says the MIL can be used for this. Who is correct?
    A. A only
    B. B only
    C. Both A and B
    D. Neither A nor B   (F.13)

83. Technician A says that fuel injectors in MFI and SFI systems are subjected to more heat than TBI injectors. Technician B says that TBI injectors have more problems with tip deposits. Who is right?
    A. Technician A only
    B. Technician B only
    C. Both A and B
    D. Neither A nor B   (D.3 and D.11)

84. During an emission test, a vehicle emits high HC at idle; but CO is too low to measure. O_2 sensor voltage is below 250 millivolts. When propane is used to richen the mixture, CO rises to measurable levels and HC drops slightly. O_2 sensor voltage increases to about 850 millivolts. What MOST likely caused the original HC readings and low O_2 voltage?
    A. High carburetor float level on a carbureted engine
    B. Leaking brake booster diaphragm
    C. Dirty air filter element
    D. High fuel pressure on a fuel-injected engine.   (E.5, E.10, F.4 and F.6)

85. Technician A says low resistance in a component can cause system problems because it may cause a high current condition. Technician B says high resistance in a circuit can cause system problems because the PCM uses low-current components. Who is right?
    A. Technician A only
    B. Technician B only
    C. Both A and B
    D. Neither A nor B   (B.14 and B.20)

86. A PFI-equipped engine is very difficult to start when the temperature is below 40°F. After completing a number of diagnostic tests, the technician has determined that the cold-start injector is not operating although there is fuel pressure present. What should the technician proceed to next?
    A. Check engine block ground.
    B. Replace oxygen sensor.
    C. Check thermo-time switch operation.
    D. Check accuracy of ECT signal.   (D.9)

87. Lean surge at low speeds and hard starting are experienced on the composite vehicle. Compression, vacuum, and electrical faults have all been ruled out. What test should the technician do next to obtain a correct fuel system diagnosis?
    A. Injector-balance test
    B. Injector-flow test
    C. Fuel-pressure test
    D. Injector-sound test   (D.9 and D.10)

88. A P0172 DTC (fuel-trim rich, bank 1) from the composite vehicle could be caused by which of the following?
    A. A leaking fuel injector
    B. Intake manifold leak
    C. EGR valve stuck closed
    D. Low charging system voltage   (D.3, D.10 and D.12)

89. What is the purpose of coating the back of an ignition module with di-electric (silicone grease) before installation?
    A. To insulate the module from the engine heat
    B. To electrically insulate the module from ground
    C. To ensure electrical ground
    D. To help dissipate heat from the module   (C.7 and C.8)

90. Technician A says a linear EGR valve can be tested using a scan tool. Technician B says a positive back pressure EGR valve can be tested with a hand-held vacuum pump and light air pressure applied to the exhaust inlet. Who is right?
    A. Technician A only
    B. Technician B only
    C. Both A and B
    D. Neither A nor B   (E.5 and E7)

Engine Speed	Idle	2000 RPM
HC (ppm)	50	150
CO (percent)	0.3	6
CO_2 (percent)	14	12
O_2 (percent)	0.3	0.2

91. Refer to the emissions chart. Based on the readings, the condition is due to:
    A. an ignition misfire.
    B. low fuel pressure at cruise.
    C. a vacuum leak at idle.
    D. over-rich at cruise.   (F.4 and F.6)

92. Technician A says that lower-than-normal CO readings can result from the EVAP system purging at idle. Technician B says higher-than-normal CO readings can result from lower-than-specified fuel pressure. Who is right?
    A. Technician A only
    B. Technician B only
    C. Both A and B
    D. Neither A nor B   (F.6, F.8 and F.10)

93. Higher-than-normal CO emissions can result from which of the following conditions?
    A. Higher-than-normal combustion chamber temperatures
    B. Leak air/fuel mixture
    C. Restricted air intake
    D. Cylinder misfiring   (D.9)

94. All of the following are symptoms that could indicate a stuck, closed, or blocked PCV valve EXCEPT:
    A. oil leaks at engine gaskets.
    B. lower-than-normal idle speed.
    C. oil in the air cleaner housing.
    D. higher-than-normal idle speed.   (E.7, E.8 and E.9)

95. A vehicle fails an I/M240 emission test for excessive HC. The drive trace shows high HC throughout the test with normal CO output that is well below the cut point. Technician A says that a charcoal canister that is purging all the time could be the cause. Technician B says the car may have a fuel line or tank leak. Who is right?
    A. Technician A only
    B. Technician B only
    C. Both A and B
    D. Neither A nor B   (E.7, E.8, F.4, F.6 and F.9)

SCAN TOOL DATA

Engine Coolant Temperature (ECT) Sensor 68°F/ 20°C/ 2.93 volts	Intake Air Temperature (IAT) Sensor 68°F/ 20°C/ 2.93 volts	Mass Airflow Sensor (MAF) 2 gm/sec/0.70 volts	Throttle Actuator Control Motor (TAC) 0 percent
Throttle Position Sensor 1 (TP1) 0 percent / 4.50 volts	Throttle Position Sensor 2 (TP2) 0 percent / 50 volts	Accelerator Pedal Position Sensor 1 (APP1) 0 percent / .50 volts	Accelerator Pedal Position Sensor 2 (APP2) 0 percent / 1.50 volts
Crankshaft Position Sensor (CPS) 500 rpm	Heated Oxygen Sensor Bank 1 (HO2S 1/1) .01 volts	Heated Oxygen Sensor Bank 2 (HO2S 2/1) 0.1 volts	Heated Oxygen Sensor Post-Cat (HO2S 1/2) 0.0 volts
Battery Voltage (B+) 12.8 volts	Fuel Enable NO	EVAP Canister Vent Solenoid OFF	Valid Ignition Key NO
Vehicle Speed Sensor (VSS) 0 MPH	Open/Closed Loop OPEN	Malfunction Indicator Lamp (MIL) OFF	Ignition Timing Advance 0 BTDC
Fuel Pump Relay (FP) ON	EVAP Canister Purge Solenoid 0 percent	P/S Switch OFF	Brake Switch OFF

Measured Ignition Timing °BTDC   Base Timing 10°   Actual Timing 11°

96. The composite vehicle has a start and die condition. The data above was retrieved as the vehicle dies. Using the scan tool information, which is the MOST likely cause of the start and die condition?
    A. A faulty power steering pressure switch
    B. Low battery voltage
    C. A faulty crankshaft position sensor
    D. A faulty ignition key   (B.2 and B.12)

97. A vehicle with sequential fuel injection (SFI) has equally high long-term fuel trim both at idle and at 3,000 rpm. Technician A says a vacuum leak may be the cause. Technician B says a dirty fuel filter or a weak fuel pump may be the cause. Who is right?
    A. Technician A only
    B. Technician B only
    C. Both A and B
    D. Neither A nor B   (D.6, D.9 and D.10)

98. Technician A says that a faulty automatic transmission may shift at an rpm that is too low or too high for load conditions. Technician B says that an automatic transmission shifting in and out of overdrive may be perceived as an engine surge. Who is right?
    A. Technician A only
    B. Technician B only
    C. Both A and B
    D. Neither A nor B

    (A.10)

7 Appendices

Answers to the Test Questions for the Sample Test Section 5

1.	C	23.	B	45.	A	67.	C
2.	D	24.	D	46.	B	68.	C
3.	C	25.	A	47.	B	69.	C
4.	C	26.	C	48.	B	70.	C
5.	D	27.	D	49.	C	71.	A
6.	A	28.	C	50.	D	72.	D
7.	C	29.	A	51.	C	73.	C
8.	C	30.	D	52.	C	74.	B
9.	C	31.	C	53.	C	75.	A
10.	B	32.	B	54.	B	76.	B
11.	C	33.	C	55.	A	77.	A
12.	C	34.	C	56.	B	78.	C
13.	C	35.	B	57.	B	79.	B
14.	D	36.	D	58.	B	80.	D
15.	B	37.	C	59.	C	81.	D
16.	C	38.	B	60.	A	82.	D
17.	A	39.	D	61.	B	83.	B
18.	B	40.	C	62.	A	84.	D
19.	D	41.	A	63.	C	85.	D
20.	C	42.	B	64.	A	86.	D
21.	C	43.	C	65.	C	87.	B
22.	C	44.	A	66.	A	88.	A

89.	A	91.	D	93.	A
90.	A	92.	B		

Explanations to the Answers for the Sample Test Section 5

Question #1
Answer A is wrong. Technician B is also correct.
Answer B is wrong. Technician A is also correct.
Answer C is correct. Both Technicians are correct.
Answer D is wrong. Both Technicians are correct.

Question #2
Answer A is wrong. An incorrect reading at the MAF would not limit the speed.
Answer B is wrong. The VSS is showing the correct MPH.
Answer C is wrong. If the Throttle Actuator were stuck the command would be at 100%.
Answer D is correct. Throttle Position Sensor 1 is not showing the correct throttle opening. When either TP sensor fails, the ECM will turn on the MIL and limit the maximum throttle opening to 35%.

Question #3
Answer A is wrong. Reference voltage is five volts and would be stable even in a low charge condition. More importantly, various sensor readings do not indicate reference voltage problems.
Answer B is Wrong. Oxygen sensor contamination would not cause this condition.
Answer C is correct. The charging system data show an undercharging condition
Answer D is wrong. A faulty starter could not cause all of the symptoms listed.

Question #4
Answer A is wrong. The TAC value is correct for this idle condition. The rpm is correct.
Answer B is Wrong. The TPS voltages are correct at the displayed value for a closed throttle position at idle.
Answer C is correct. The MAF sensor is sending a key-on/engine-off signal while the engine is at idle. This reading can cause false fuel calculations resulting in a hard-start or poor performance.
Answer D is wrong. There is no indication of a fuel pressure problem, as indicated by the oxygen sensors.

Question #5
Answer A is wrong. Failed power and grounds would not explain the seasonal nature of the problem.
Answer B is Wrong. Insufficient fuel volume would cause a constant condition.
Answer C is wrong. An ignition coil problem would not explain the seasonal nature of the condition.
Answer D is correct. If the thermostat is stuck open, the engine would run too cool in the winter, and reduced power and fuel economy would result.

Question #6
Answer A is correct. Only Technician A is correct. No crankshaft signal will cause the PCM to have no reference and, therefore, no injection pulse or ignition operation.
Answer B is Wrong. A TPS stuck at WOT would cut off fuel, not the spark.
Answer C is wrong. Only Technician A is correct.
Answer D is wrong. Only Technician A is correct.

Question #7
Answer A is wrong. Technician B is also correct.
Answer B is wrong. Technician A is also correct.
Answer C is correct. Both Technicians are correct.
Answer D is wrong. Both Technicians are correct.

Question #8

Answer A is wrong. If the converter were working normally, the trace of the downstream sensor would be smoother.
Answer B is wrong. Because the sensor is responding, the sensor has not failed.
Answer C is correct. If the traces are the same, no change in the exhaust gasses is taking place in the converter.
Answer D is wrong. The downstream trace should be smoother if the converter is doing its job. An OBD II vehicle monitors the operation of the catalytic converter by comparing the precatalytic O_2 sensor with the postcatalytic O_2 sensor. If the catalytic converter is functioning properly, the catalytic converter will give up its stored oxygen causing an almost flat line signal from the post O_2 sensor. In the event the catalytic converter fails, the post O_2 sensor will almost menace the pre O_2 sensor.

Question #9

Answer A is wrong. Technician B is also correct.
Answer B is wrong. Technician A is also correct.
Answer C is correct. Both Technicians are correct.
Answer D is wrong. Both Technicians are correct.

Question #10

Answer A is wrong. SAE communications classification began with Class A being the slowest, Class B being medium speed and Class C the fastest.
Answer B is correct. Only Technician B is correct. Multiplexing allows the use of the same sensor by multiple computers with each computer being hardwired to it.
Answer C is wrong. Only Technician B is correct.
Answer D is wrong. Only Technician B is correct.

Question #11

Answer A is wrong. Technician B is also correct.
Answer B is wrong. Technician A is also correct.
Answer C is correct. Both Technicians are correct. The scan tool is used to check for software versions of the PCM and DLC is the connection used for flash programming.
Answer D is wrong. Both Technicians are correct.

Question #12

Answer A is wrong. A zero to 2-Hertz frequency is too slow for normal oxygen sensor operation at 200 rpm.
Answer B is Wrong. A 5 to 25-Hertz frequency is too fast and would indicate probably engine misfire conditions.
Answer C is correct. A 0.5 to 5-Hertz frequency is the established normal value for oxygen sensor frequency at 2500 rpm.
Answer D is wrong. A frequency over 5-Hertz is too fast and indicates a fuel control problem.

Question #13

Answer A is wrong. Technician B is also correct.
Answer B is wrong. Technician A is also correct.
Answer C is correct. Both Technicians are correct.
Answer D is wrong. Both Technicians are correct.

Question #14

Answer A is wrong. Major components can only be replaced with manufacturer's specified parts, or equivalent.
Answer B is wrong. Under Federal Law, emission components cannot be bypassed
Answer C is wrong. Neither Technician is correct.
Answer D is correct. Neither Technician is correct.

Question #15
Answer A is wrong. Similarities aside, the technician has no way of knowing what changes may have been made to the system and its related diagnostic procedures.
Answer B is correct. Only Technician B is correct. Only with the proper diagnostic information can the technician be sure of using the correct procedures.
Answer C is wrong. Only Technician B is correct.
Answer D is wrong. Only Technician B is correct.

Question #16
Answer A is wrong. Technician B is also correct.
Answer B is Wrong. Technician A is also correct.
Answer C is correct. Both Technicians are correct. Most OBD II compliant vehicles test the EVAP system for purge flow when commanded by the PCM and will also test the system for leaks if equipped with an enhanced EVAP system.
Answer D is wrong. Both Technicians are correct.
Since 1996, some vehicles have an enhanced evaporative system monitor. This system detects leaks and restrictions in the EVAP system. A newly designed fuel tank cap is used on this system. In these enhanced systems, an evaporative system leak or a missing fuel tank cap will cause the MIL to turn on.

Question #17
Answer A is correct. Only Technician A is correct. Always verify and clarify the customer complaint as the first step in a diagnosis.
Answer B is Wrong. The correct, or relevant, diagnostic information can only be identified after the complaint is verified.
Answer C is wrong. Only Technician A is correct.
Answer D is wrong. Only Technician A is correct. Typical service procedures consist of verifying the complaint, doing a good visual inspection, retrieving DTCs , checking for any TSBs, looking at the scan tool data, repairing the fault and test-driving the vehicle afterward.

Question #18
Answer A is wrong. Block-learn (long-term fuel trim) numbers above 128 indicate a rich correction for a lean-exhaust condition.
Answer B is correct. The PCM is correcting for an overly lean exhaust condition. Answer A is wrong. Block-learn (long-term fuel trim) numbers above 128 indicate a rich correction for a lean-exhaust condition.
Answer C is wrong. Ignition timing would not influence block-learn numbers to this extent.
Answer C is wrong. Ignition timing would not influence block-learn numbers to this extent.Block-learn (long-term fuel trim) is based on the feedback from the short-term strategies. (Integrator) Short-term changes are not saved in the computer's memory. All changes to the fuel system happen immediately and occur in direct response to the O_2 sensor and other sensors. Long-term changes are saved in the computer's memory. These stored values are used the next time the engine is operated in a similar situation and under similar conditions. No adjustment is seen as 0% or 128 as the computer adds fuel the percentage goes up or the number increases from 128, as the computer decreases fuel the percentage goes down or the number decreases from 128.

Question #19
Answer A is wrong. The microprocessor is the decision-making chip in the computer.
Answer B is Wrong. Signal conditioning involves amplification and analog-to-digital conversion of input signals.
Answer C is wrong. Readout controls are fictitious.
Answer D is correct. The read-only memory (ROM) contains lookup and calibration tables.
The microprocessor can read information from the ROM, but information cannot be written into the ROM by the microprocessor and the microprocessor cannot erase ROM information. The ROM contains lookup tables that contain information about how a vehicle should perform. For example, the look-up table would contain the ideal manifold vacuum under various engine-operating conditions.

Question #20
Answer A is wrong. Technician B is also correct.
Answer B is wrong. Technician A is also correct.
Answer C is correct. Both Technicians are correct. The enabling criteria must be met in order to duplicate the conditions when the fault was set. The drive cycle for the system at fault must be performed in order to duplicate the fault.
Answer D is wrong. Both Technicians are correct.

Question #21
Answer A is wrong. Technician A is correct that a scan tool enables a technician to read DTCs.
Answer B is Wrong. Technician B is correct that a scan tool allows a technician to compare vehicle data readings to each other and to manufacturer's specifications.
Answer C is correct. Both Technicians are correct.
Answer D is wrong. Both Technicians are correct.

Question #22
Answer A is wrong. Technician B is also correct.
Answer B is wrong. Technician A is also correct.
Answer C is correct. Both Technicians are correct.
Answer D is wrong. Both Technicians are correct.

Question #23
Answer A is wrong. The radio can be reprogrammed easily.
Answer B is correct. Unregulated alternator output voltage above 16 volts can damage onboard computers.
Answer C is wrong. Loss of this information will not damage the vehicle or its systems.
Answer D is wrong. Adaptive strategy can be relearned relatively easily.

Question #24
Answer A is wrong. A slipped timing belt affects all cylinders.
Answer B is Wrong. A broken timing chain affects all cylinders.
Answer C is wrong. A leaking intake valve affects compression but would not reduce airflow.
Answer D is correct. A worn intake camshaft lobe will reduce the valve opening and therefore reduce airflow to that cylinder.

Question #25
Answer A is correct. Only Technician A is correct. The best place to start is with a service manual and DMM.
Answer B is wrong. Only Technician A is correct.
Answer C is wrong. Only Technician A is correct.
Answer D is wrong. Only Technician A is correct.

Question #26
Answer A is wrong. Technician B is also correct.
Answer B is wrong. Technician A is also correct.
Answer C is correct. Both Technicians are correct.
Answer D is wrong. Both Technicians are correct.

Question #27
Answer A is wrong. A properly functioning pump would have a high-voltage drop.
Answer B is Wrong. A high-voltage drop across the switch would mean the switch is faulty.
Answer C is wrong. Neither Technician is correct.
Answer D is correct. Neither Technician is correct.

Question #28

Answer A is wrong. Excessive air inlet temperature may cause a very slight decrease in CO_2, but would more likely have no effect at all.

Answer B is Wrong. A lean air/fuel mixture would only cause a very slight decrease in CO_2.

Answer C is correct. Extremely retarded timing decreases combustion time, which would significantly lower CO_2 emissions.

Answer D is wrong. Dirty fuel injectors can result in a lean air/fuel mixture, which would only have a slight effect on CO_2 emissions.

Question #29

Answer A is correct. Only Technician A is correct.

Answer B is wrong. Only Technician A is correct.

Answer C is wrong. Only Technician A is correct.

Answer D is wrong. Only Technician A is correct.

Question #30

Answer A is wrong. The pre-HO_2S waveform is normal.

Answer B is Wrong. The post-HO_2S waveform is showing activity within its operating range.

Answer C is wrong. The pre-HO_2S does not show an overly rich mixture.

Answer D is correct. The switching frequency of the pre- and post-oxygen sensors are almost the same indicating a failing catalytic converter. The post-HO_2S should show far fewer cycles than the pre-HO_2S when the converter is working normally. If the converter is operating properly, the signal from the precatalyst O_2 will have oscillations while the post catalyst O_2 will be relatively flat.

Question #31

Answer A is wrong. Excessive fuel pressure would not produce the elevated combustion chamber temperatures that increase NO_x production.

Answer B is Wrong. Lower coolant temperatures would not produce the elevated combustion chamber temperatures that increase NO_x production.

Answer C is correct. Carbon buildup in the combustion chamber would raise compression, leading to higher combustion chamber temperatures and increased NO_x formation.

Answer D is wrong. A restricted air filter will not cause high NO_x emissions.

Question #32

Answer A is wrong. Only Technician B is correct. The PCM will not turn on the fuel pump relay until it sees a crank signal from the crankshaft position sensor. You need to remember that this is a command and not the result.

Answer B is correct. Only Technician B is correct. There should be an RPM reading during cranking with a functional Crankshaft Position Sensor.

Answer C is wrong. Only Technician B is correct.

Answer D is wrong. Only Technician B is correct.

Question #33

Answer A is wrong. Technician B is also correct.

Answer B is Wrong. Technician A is also correct.

Answer C is correct. Both Technicians are correct. An O_2 sensor reads oxygen content in the exhaust; if cylinder compression is low there won't be complete burning of the gases, one of which is oxygen. This unused oxygen will cause the O_2 sensor to read high oxygen levels. The same is true of a vacuum leak. Any illegal air entering the engine shows up as high oxygen level in the exhaust.

Answer D is wrong. Both Technicians are correct.

Question #34
Answer A is wrong. Technician B is also correct.
Answer B is Wrong. Technician A is also correct.
Answer C is correct. Both Technicians are correct. Never use a test light or DVOM with low impedance on any computer circuit unless specified by the service manual. Use of a test light or DVOM with low impedance will draw a lot of amperage and could damage computer circuits. It is also very important to use proper jump stating techniques when jump stating a vehicle to prevent damage to the computer circuits.
Answer D is wrong. Both Technicians are correct.

Question #35
Answer A is wrong. A restricted fuel filter would cause low fuel pressure.
Answer B is correct. A restricted fuel-pressure regulator would cause high fuel pressure. The way the pressure regulator works is by restricting the return flow of fuel so a restricted pressure regulator would increase fuel pressure.
Answer C is wrong. Low voltage to the fuel pump would cause low fuel pressure.
Answer D is wrong. A poor ground would cause low fuel pressure.

Question #36
Answer A is wrong. High combustion chamber temperature and a lean air/fuel mixture increase NOx emissions and do not mask the condition.
Answer B is Wrong. A lean air/fuel mixture increases NOx emissions and does not mask the condition.
Answer C is wrong. A rich mixture does not have a heating effect on the combustion chambers.
Answer D is correct. A rich air/fuel mixture often increases CO emissions, and its cooling effect can mask a NOx emission problem. Correcting the air/fuel ratio can reduce CO emissions, but then the NOx problem can become evident.

Question #37
Answer A is wrong. Rich-to-lean transition should be less than 125 ms.
Answer B is Wrong. Rich-to-lean transition should be less than 125 ms.
Answer C is correct. This demonstrates the integrity of the O_2 sensor. Slower switching could mean a worn sensor.
Answer D is wrong. Rich-to-lean transition could not occur in 10 ms.

Question #38
Answer A is wrong. A crankshaft position sensor would more than likely affect all cylinders and is not associated with secondary ignition circuit.
Answer B is correct. An open in the number two spark plug wire would cause a higher firing voltage requirement. This open would not leave enough voltage to fire the spark plug properly, thus a noticeable misfire under load and higher HC due to incomplete combustion.
Answer C is wrong. Low compression would not cause a high firing spike it would cause the internal combustion chamber resistance to go down resulting in a low firing line.
Answer D is wrong. If number two spark plug were shorted to ground, the firing voltage would be lower than the rest.

Question #39
Answer A is wrong. Use of improper transmission fluid can cause drivability problems in many late model transmissions such as shudder when the torque converter clutch applies or harsh shifting.
Answer B is Wrong. DEXRON-III® is used in many late-model transmissions, but it specifically should not be used in some vehicles.
Answer C is wrong. Neither Technician is correct.
Answer D is wrong. Neither Technician is correct.

Question #40

Answer A is wrong. Technician A is partially right. He should also check injector resistance to find the cause of the injector driver failure.

Answer B is Wrong. Technician B is partially right. After checking the injector resistance and correcting the problem, he should then replace the PCM.

Answer C is correct. Both Technicians are correct.

Answer D is wrong. Both Technicians are correct.

Question #41

Answer A is correct. The scan tool data shows a charging voltage of 16.8 volts, which could damage the ECM or components.

Answer B is Wrong. Although a bad ground could cause a problem, the data does not show this.

Answer C is wrong. The charging data shows an overcharge condition.

Answer D is wrong. TPS voltages are normal.

Question #42

Answer A is wrong. No emission test requires measurement of emissions upstream from the catalytic converter.

Answer B is correct. Only Technician B is correct. The measurement units for I/M testing are grams per mile (gpm).

Answer C is wrong. Only Technician B is correct.

Answer D is wrong. Only Technician B is correct.

The I/M240 (Inspection/Maintenance 240 seconds) test requires the use of a chassis dynamometer, commonly called a dyno. While on the dyno, the vehicle is operated for 240 seconds and under different load conditions. The test drive on the dyno simulates both in-traffic and highway driving and stopping. The emission tester tracks the exhaust quality through these conditions in grams per mile.

Question #43

Answer A is wrong. Serial describes a data format used by scan tools.

Answer B is Wrong. A hall-effect switch does not produce an analog signal.

Answer C is correct. The high-low, on-off signal from a hall-effect switch is a digital signal.

Answer D is wrong. A hall effect switch does not produce an alternating current (AC) signal.

Question #44

Answer A is correct. A CKP sensor is a crankshaft position sensor. This is used for rpm and position input.

Answer B is Wrong. A TP sensor is a throttle position sensor and has nothing to do with crankshaft speed and position.

Answer C is wrong. A VS sensor is a vehicle speed sensor and has nothing to do with crankshaft speed and position.

Answer D is wrong. A CMP sensor is a crankshaft position sensor. This is used for synchronization.

Question #45

Answer A is correct. Only Technician A is correct. The test light can be used to test for available voltage and to check for primary triggering. There should be battery voltage minus the voltage drop of the ignition resistor (if used).

Answer B is Wrong. Self-powered test lamps are used to check continuity, not circuit function.

Answer C is wrong. Only Technician A is correct.

Answer D is wrong. Only Technician A is correct.

Question #46

Answer A is wrong. Some vehicles were partially compliant with OBD II, but they were not required to be.

Answer B is correct. Full OBD II compliance was mandated for 1996.

Answer C is wrong. By 1996, all passenger vehicles were to comply fully with OBD II.

Answer D is wrong. Some vehicles were partially compliant with OBD II, but they were not required to be.

Question #47
Answer A is wrong. Replacing the ICM at this point is extremely premature.
Answer B is correct. Check the battery-voltage supply for the circuit before testing any individual parts.
Answer C is wrong. The secondary circuit should not affect the primary ignition circuit operation. Additionally, see the explanation for answer B.
Answer D is wrong. See the explanation for answer B.

Question #48
Answer A is wrong. Only Technician B is correct. A faulty IAC motor would not be the root cause of this problem.
Answer B is correct. Only Technician B is correct. If the air-injection system continuously pumps air upstream into the exhaust, the O_2 sensor will measure a continuously lean exhaust. As a result, the PCM will enrich the air/fuel mixture, which leads to poor gas mileage and an overloaded catalytic converter and resulting smell.
Answer C is wrong. Only Technician B is correct.
Answer D is wrong. Only Technician B is correct. This is a common fault of the air/switching valve. When an engine is started with a secondary air injection system the air is injected upstream in the exhaust manifold usually. After an elapsed time (usually closed loop) the air is switched downstream to the catalytic converter. In the event the air does not switch downstream it floods the O_2 sensor with oxygen causing the computer to drive the system rich, resulting in poor fuel economy and a foul smell from the exhaust.

Question #49
Answer A is wrong. Technician B is also correct.
Answer B is Wrong. Technician A is also correct.
Answer C is correct. Both Technicians are correct. Although there may be a code in the PCM, the MIL will not come on until the failure repeats during the second trip. This is defined in the composite vehicle information.
Answer D is wrong. Both Technicians are correct.

Question #50
Answer A is wrong. This condition would result in elevated HC and CO emissions.
Answer B is Wrong. This condition would result in elevated HC and CO emissions.
Answer C is wrong. This condition would result in elevated HC and CO emissions.
Answer D is correct. The secondary air injection systems function is to supply additional air to the oxidation catalyst to help in the converting of HC and CO to CO_2 and water. In the event that the secondary air system becomes inoperative there will be an increase in HC and CO.

Question #51
Answer A is wrong. Technician B is also correct.
Answer B is wrong. Technician A is also correct.
Answer C is correct. Both Technicians are correct.
Answer D is wrong. Both Technicians are correct.

Question #52
Answer A is wrong. Technician B is also correct.
Answer B is Wrong. Technician A is also correct.
Answer C is correct. Both Technicians are correct. Any time a rotor button fails it is good practice to check the rest of the secondary system. Many times a failed rotor button is the result of a sparkplug wire or spark plug with high resistance.
Answer D is wrong. Both Technicians are correct.

Question #53

Answer A is wrong. This indicates a pickup coil shorted to ground.

Answer B is Wrong. This indicates an internally shorted pickup coil.

Answer C is correct. The pickup coil should be isolated from ground, which is correctly indicated by infinite resistance in this measurement.

Answer D is wrong. This indicates an open pickup coil.

Pickup coils are magnetic pulse generating trigger devices that produce an A/C voltage. These sensors are used to trigger the ignition module to collapse the magnetic field in the ignition coil. Pickup coil resistance is typically 150−1500 ohms depending on the manufacturer.

Question #54

Answer A is wrong. Vehicle recall letters have no direct advantage in troubleshooting.

Answer B is correct. Vehicle recall letters have no direct advantage in troubleshooting.

Answer C is wrong. Service manuals are necessary for proper diagnosis.

Answer D is wrong. Service bulletins are often necessary for proper diagnosis.

Question #55

Answer A is correct. Only Technician A is correct. The range of correction that is programmed into the operating strategy is designed to correct for misadjustments or natural engine wear and other known variables. It is not designed, however, to be a substitute for proper maintenance or repairs.

Answer B is Wrong. A scan tool cannot be used to test compression. However, certain vehicles allow certain scanners to perform a cylinder power balance test-not a compression test.

Answer C is wrong. Only Technician A is correct.

Answer D is wrong. Only Technician A is correct. Block-learn (long-term fuel trim) is based on the feedback from the short-term strategies. (Integrator) short-term changes are not saved in the computers memory. All changes to the fuel system happen immediately and occur in direct response to the O_2 sensor and other sensors. Long-term changes are saved in the computer's memory. These stored values are used the next time the engine is operated in a similar situation and under similar conditions. No adjustment is seen as 0% or 128 as the computer adds fuel the percentage goes up or the number increases from 128, as the computer decreases fuel the percentage goes down or the number decreases from 128.

Question #56

Answer A is wrong. "Monitor sensor data for invalid inputs" is a meaningless generalization.

Answer B is correct. Low compression is a common cause of high HC emissions under all driving conditions, as these test results indicate.

Answer C is wrong. An injector balance test would be unlikely to pinpoint the cause of high HC emissions under all driving conditions.

Answer D is wrong. High fuel pressure would cause increased CO emissions, along with higher than normal HC.

Question #57

Answer A is wrong. The lock-up clutch could only cause stalling if it operated but did not release.

Answer B is correct. Only Technician B is correct. If the torque converter clutch does not engage, emissions may be higher than normal at cruising speed, and there will be lower fuel economy.

Answer C is wrong. Only Technician B is correct.

Answer D is wrong. Only Technician B is correct.

An inoperative torque converter clutch would result in torque multiplication at on times, which would increase emissions and decrease fuel economy. Once the torque converter goes in to lockup, the turbine is locked to the converter cover that is bolted to the flywheel; this eliminates any torque multiplication.

Question #58

Answer A is wrong. It is more efficient to check connections at the sensors before checking connections at the PCM.
Answer B is correct. It is more efficient to check connections at the sensors before checking connections at the PCM.
Answer C is wrong. Measuring magnetic strength is not a valid test.
Answer D is wrong. Magnetic timing offset is an ignition timing procedure that is not related to CKP or CMP sensor troubleshooting.

Question #59

Answer A is wrong. There is no way of determining if the injector is open too long because engine operating conditions are not stated in the question.
Answer B is Wrong. The waveform indicates supply voltage is very near normal charging system voltage.
Answer C is correct. The low voltage spike when the injector turns off indicates a shorted injector winding. Normal voltage spike levels are between 60 to over 100 volts.
Answer D is wrong. This low injector spike is not normal operation.

Question #60

Answer A is correct. Only Technician A is correct.
Answer B is wrong. Only Technician A is correct.
Answer C is wrong. Only Technician A is correct.
Answer D is wrong. Only Technician A is correct.

Question #61

Answer A is wrong. The engine stays in open loop while warming up.
Answer B is correct. Only Technician B is correct. Requirements of temperature, time, and O_2 switching must be met before switching to closed loop. The vehicle operates in open loop until the oxygen sensor reaches a certain temperature of approximately 600°F, a certain amount of time has elapsed and the engine is at a certain temperature.
Answer C is wrong. Only Technician B is correct.
Answer D is wrong. Only Technician B is correct.

Question #62

Answer A is correct. Service intervals for emission control devices are listed in the maintenance schedules found in owners' manuals. Owner's manuals contain information on operating the controls of a vehicle as well as maintenance information such as lubrication type as well as service intervals.
Answer B is Wrong. Parts-location books do not contain service interval information. Parts location books show exact location of components on the vehicle.
Answer C is wrong. Vacuum hose diagrams do not contain service interval information. Vacuum hose diagrams are used for proper vacuum hose routing of a vehicle.
Answer D is wrong. Diagnostic procedures do not contain service interval information. Diagnostic procedures are step-by-step procedures used for diagnosing system or circuit failures.

Question #63

Answer A is wrong. Cutpoints describe the thresholds for emission test failure.
Answer B is Wrong. Transient describes a condition in the drive cycle test.
Answer C is correct. Most emission readings recorded by the exhaust analyzer always lag behind the driving event that caused the change in emissions.
Answer D is wrong. Deceleration enrichment has nothing to do with the traces shown here.

Question #64

Answer A is correct. Only Technician A is correct. A continuous low O_2 sensor voltage indicates a lean exhaust condition. Injector pulse width would increase to richen the intake air/fuel mixture.

Answer B is Wrong. Block-learn numbers (long-term fuel trim) would increase to richen the intake air/fuel mixture.

Answer C is wrong. Only Technician A is correct.

Answer D is wrong. Only Technician A is correct.Block-learn (long-term fuel trim) is based on the feedback from the short-term strategies. (Integrator) short-term changes are not saved in the computer's memory. All changes to the fuel system happen immediately and occur in direct response to the O_2 sensor and other sensors. Long-term changes are saved in the computer's memory. These stored values are used the next time the engine is operated in a similar situation and under similar conditions. No adjustment is seen as 0% or 128 as the computer adds fuel the percentage goes up or the number increases from 128, as the computer decreases fuel the percentage goes down or the number decreases from 128.

Question #65

Answer A is wrong. Technician B is also correct.

Answer B is wrong. Technician A is also correct.

Answer C is correct. Both Technicians are correct. Always verify the effectiveness of the repair and clear any stored diagnostic trouble codes.

Answer D is wrong. Both Technicians are correct.

Question #66

Answer A is correct. Only Technician A is correct.

Answer B is wrong. Only Technician A is correct.

Answer C is wrong. Only Technician A is correct.

Answer D is wrong. Only Technician A is correct.

Question #67

Answer A is wrong. Technician B is also correct.

Answer B is wrong. Technician A is also correct.

Answer C is correct. Both Technicians are correct.

Answer D is wrong. Both Technicians are correct.

Question #68

Answer A is wrong. Technician A is correct. A clogged EGR passage would not supply the exhaust gas to cool the combustion chamber temperatures.

Answer B is Wrong. Technician B is correct. A hole in the vacuum line would prevent the EGR valve from actuating.

Answer C is correct. Both Technicians are correct.

Answer D is wrong. Both Technicians are correct. The sole purpose of the EGR system is to lower the combustion chamber temperature; this is done to lower the production of NOx which is produced under high temperature conditions.

Question #69

Answer A is wrong. An EGR valve that does not open enough leads to higher than normal combustion chamber temperatures, not lower temperatures.

Answer B is Wrong. An EGR valve that does not open enough leads to higher NO_x emissions, not lower.

Answer C is correct. An EGR valve that does not open enough leads to higher NO_x emissions.

Answer D is wrong. This EGR valve condition would not cause a misfire.The sole purpose of the EGR system is to lower the combustion chamber temperature; this is done to lower the production of NO_x which is produced under high temperature conditions.

Question #70

Answer A is wrong. High levels of alcohol in the fuel system can result in corrosion. Alcohol (methanol) is corrosive to lead, aluminum, plastic and rubber. It can cause rubber to swell.

Answer B is Wrong. High levels of alcohol can result in a no-start condition. Alcohol (ethanol) raises the volatility of gasoline about 0.5 psi, which can cause hot drivability or even a no start in hot temperatures.

Answer C is correct. Alcohol does not enrich the air/fuel mixture.

Answer D is wrong. High levels of alcohol in the fuel system can result in fuel filter plugging. Alcohol cleans the fuel system, which can cause clogged fuel filters.

Question #71

Answer A is correct. Only Technician A is correct. As an (EGR) exhaust gas recirculation valve becomes carboned up it may begin to stick. The best repair is to replace the valve.

Answer B is Wrong. Attempting to clean a dirty (EGR) exhaust gas recirculation valve will usually result in a recheck and a dissatisfied customer. The best repair is to replace.

Answer C is wrong. Only Technician A is correct.

Answer D is wrong. Only Technician A is correct.

Question #72

Answer A is wrong. Additional fuel from a leaking cold-start injector can cause a warm engine to stall from an overly rich mixture.

Answer B is Wrong. Additional fuel from a leaking cold-start injector can cause a warm engine to idle roughly from an overly rich mixture.

Answer C is wrong. Additional fuel from a leaking cold-start injector can cause hard-starting problems with a warm engine.

Answer D is correct. The rich mixture created by a leaking cold-start injector would tend to reduce combustion temperature, which works against detonation. Detonation is usually associated with high temperature; a leaking cold start injector would cause a lower temperature due to the over rich condition.

Question #73

Answer A is wrong. High fuel pressure would not cause the symptoms listed in the question and would not cause the problem to disappear when the engine sat for a short time.

Answer B is wrong. A weak ignition coil would not cause the symptoms to come and go.

Answer C is correct. A malfunctioning heated air intake could allow throttle body icing to occur and cut-off airflow to the engine creating the stalling problem. With the engine stopped, heat would melt the ice and the engine would work properly again for a short time.

Answer D is wrong. Over-advanced ignition timing would not cause the symptoms listed in the question. A throttle body or carburetor can ice up in temperatures as high as 50 degrees due to the velocity of the airflow. Icing is more likely during high humidity conditions. One of the roles of the heated inlet system is to prevent icing.

Question #74

Answer A is wrong. A pinched or restricted fuel return line would increase fuel pressure, not lower fuel pressure.

Answer B is correct. Only Technician B is correct. Low supply voltage to the electric fuel pump can lower the pump speed and reduce pressure and volume output.

Answer C is wrong. Only Technician B is correct.

Answer D is wrong. Only Technician B is correct.

Question #75

Answer A is correct. Only Technician A is correct. Use of a lab scope is one of the best ways to diagnose a bad oxygen sensor. A DVOM may also be used to monitor voltage.

Answer B is Wrong. The oxygen sensor heater can be checked for resistance, but the signal wire and ground can not.

Answer C is wrong. Only Technician A is correct.

Answer D is wrong. Only Technician A is correct.

Question #76
Answer A is wrong. Only Technician B is correct.
Answer B is correct. Only Technician B is correct.
Answer C is wrong. Only Technician B is correct.
Answer D is wrong. Only Technician B is correct.

Question #77
Answer A is correct. Only Technician A is correct. Technician A has identified the short in the power steering pressure switch circuit. While running with the steering wheel straight ahead, there is no excessive power steering pressure; therefore, the switch would be open. With pressure, the switch closes. The PCM will interrupt the current to the A/C clutch and increase the idle, anticipating the additional load from the power steering.
Answer B is Wrong. The switch in the circuit of PCM terminal 65 is the A/C request. It would be closed when the A/C is turned on.
Answer C is wrong. Only Technician A is correct.
Answer D is wrong. Only Technician A is correct.

Question #78
Answer A is wrong. Plugged or restricted fuel-return lines can increase fuel pressure.
Answer B is Wrong. A sticking fuel-pressure regulator can increase fuel pressure.
Answer C is correct. This would cause the pump to run slow, thus reducing fuel pressure.
Answer D is wrong. A leaking vacuum hose at the pressure regulator will increase fuel pressure.
Fuel pressure is regulated by restricting the flow of fuel at the regulator. The regulator is located in the fuel return side of the system. Any restriction in the return side would increase fuel pressure. The pressure regulator in a SFI system has a vacuum hose going to it that reduces pressure at idle, but allows the pressure to increase slightly at hard throttle or WOT so the multiple injectors do not cause a pressure drop in the system. A vacuum leak at the pressure regulator hose would cause an increase in pressure.

Question #79
Answer A is wrong. Worn piston rings reduce compression, not airflow.
Answer B is correct. Worn camshaft lobes reduce valve duration and thus airflow into the affected cylinders.
Answer C is wrong. Worn valve guides reduce oil control.
Answer D is wrong. Worn cylinder walls reduce compression, not airflow.
Worn camshaft lobes is the only one of the four listed that would affect airflow. Worn camshaft lobes can be very hard to diagnose by conventional methods. A cranking compression test usually won't show this condition. A running compression test is very effective in diagnosing a worn camshaft.

Question #80
Answer A is wrong. The greater the pressure drop, the more the fuel quantity passing through the injector.
Answer B is Wrong. The smaller the pressure drop, the lower the fuel quantity passing through the injector.
Answer C is wrong. Neither Technician is correct.
Answer D is correct. Neither Technician is correct.The pressure drop test involves using a special injector pulsing tool to pulse each injector the exact amount. The fuel system is energized, the pressure is recorded, the injector is pulsed and fuel pressure after the pulsing is recorded. This is done for each injector; then pressure drops of all the injectors are compared. Most manufacturers recommend the pressures be within 1.5 psi (10 kPa) of each other.

Question #81
Answer A is wrong. PCV valves and connections should be included in a preliminary inspection.
Answer B is Wrong. Vacuum hose connections should be included in a preliminary inspection.
Answer C is wrong. Catalytic converters and exhaust systems should be inspected for missing components or leaking connections in a preliminary inspection.
Answer D is correct. Testing catalytic converter inlet and outlet temperatures is a specialized test that requires special test equipment. Choices A, B, and C are all done by immediate visual inspection: thus preliminary.

Question #82
Answer A is wrong. Voltmeters used in automotive service cannot measure high secondary voltages.
Answer B is Wrong. Ohmmeters measure resistance, not voltage.
Answer C is wrong. An ignition module tester does not check the ignition coil; it tests the primary circuit in the module.
Answer D is correct. Connecting an approved test spark plug to the coil will indicate whether or not the coil can deliver secondary voltage sufficient to jump the plug gap.

Question #83
Answer A is wrong. Low compression can cause a misfire.
Answer B is correct. Carbon deposits will not usually cause a continuous misfire.
Answer C is wrong. Intake vacuum leaks can cause a misfire.
Answer D is wrong. Worn distributor bushings can cause a misfire.

Question #84
Answer A is wrong. Hard starting is a symptom of the EGR valve stuck open.
Answer B is Wrong. Rough idle is a symptom of the EGR valve stuck open.
Answer C is wrong. Engine stalling is a symptom of the EGR valve stuck open. The sole purpose of the EGR system is to lower the combustion chamber temperature; this is done to lower the production of NOx which is produced under the high temperature conditions. With a stuck-closed EGR valve, the cylinder would overheat and detonation could occur.
Answer D is correct. EGR is a primary method of controlling detonation in a computer-controlled engine. If the valve is stuck closed, detonation is more likely to occur.

Question #85
Answer A is wrong. A rich fuel mixture would not cause this problem.
Answer B is Wrong. An EGR valve stuck closed would not cause this problem.
Answer C is wrong. A clogged catalytic converter would restrict exhaust flow but would not cause this problem.
Answer D is correct. A PCV valve stuck closed would cause excessive crankcase pressure that would force crankcase vapors through the clean-air hose and soak the air filter with oil.

Question #86
Answer A is wrong. Technician A is wrong. An open in the CAN could cause a fault to be generated but would not cause a no start.
Answer B is Wrong. Technician B is wrong. CAN is the newest form of communication. It is not on all OBD II vehicles.
Answer C is wrong. Neither Technician is correct.
Answer D is correct. Neither Technician is correct.

Question #87
Answer A is wrong. Parts per million (ppm) is not the measurement standard used for the I/M240 test.
Answer B is correct. Grams per mile (gpm) is the measurement standard used for the I/M240 test.
Answer C is wrong. Ratio is not a measurement for the emission test.
Answer D is wrong. Percentage is not used in the I/M240 test results.
The I/M240 (Inspection/Maintenance 240 seconds) test requires the use of a chassis dynamometer, commonly called a dyno. While on the dyno, the vehicle is operated for 240 seconds and under different load conditions. The test drive on the dyno simulates both in-traffic and highway driving and stopping. The emission tester tracks the exhaust quality through these conditions in grams per mile.

Question #88
Answer A is correct. A scan tool is not used when inspecting for tampering. Initially, inspections are done visually.
Answer B is wrong. Parts location manuals are very useful when inspecting for tampering.
Answer C is wrong. Wiring diagrams are useful when inspecting for tampering.
Answer D is wrong. Service bulletins may be useful when inspecting for tampering.

Question #89

Answer A is correct. Volatile RAM is a memory device that is erased when power is removed, such as by turning off the ignition. Timers and short term fuel trim can be among the items erased.
Answer B is Wrong. Nonvolatile RAM retains memory when the ignition is turned off.
Answer C is wrong. ROM devices retain their memory when power is removed by any means.
Answer D is wrong. PROMs retain their memory when power is removed by any means.

Question #90

Answer A is correct. TPS 1 is outside the operating parameters of 4.50 volts to .50 volts. This is generating a TPS DTC for TPS 1 causing the ECM to restrict throttle opening to 35%.
Answer B is wrong. TPS 2 value is correct for the situation.
Answer C is wrong. APP1 value is correct for the situation.
Answer D is wrong. APP2 value is correct for the situation.

Question #91

Answer A is wrong. A restricted fuel line or clogged fuel filter can cause this problem, but not on the composite vehicle.
Answer B is Wrong. A bad ignition module can cause this problem, but not on the composite vehicle.
Answer C is wrong. Neither Technician is correct.
Answer D is correct. Neither Technician is correct. This question illustrates that you must read the description of the "composite vehicle" on the L1 test very carefully. Under "Fuel Cutoff Model," the "Composite Vehicle Information" says: "The PCM will turn off the fuel injectors, for safety reasons, when the vehicle speed reaches 110 mph or if the engine speed exceeds 6000 rpm."

Question #92

Answer A is wrong. A PCM should not be replaced until all other resources and possible causes of a problem are eliminated, including manufacturer's service bulletin information.
Answer B is correct. Before placing a PCM, search any available sources of recent manufacturer's information.
Answer C is wrong. Only Technician B is corrected.
Answer D is wrong. Only Technician B is correct. Typical service procedures consist of verifying the complaint, doing a good visual inspection and simple test such as vacuum, compression, and cylinder leakage, retrieving DTCs, checking for any TSBs, looking at the scan tool data, repairing the fault, and test-driving the vehicle afterward.

Question #93

Answer A is correct. Service updates or bulletin information are published when nonstandard or field repairs are needed. It may include a revised service procedure, reprogramming requirements, or a replacement engine/lock-up calibration component in the PCM.
Answer B is Wrong. Diagnostic procedures have not condemned the PCM.
Answer C is wrong. High-octane fuel will only mask the root cause.
Answer D is wrong. Idle-speed adjustment will only mask the root cause.
Typical service procedures consist of verifying the complaint, doing a good visual inspection and simple test such as vacuum, compression, and cylinder leakage, retrieving DTCs, checking for any TSBs, looking at the scan tool data, repairing the fault, and test-driving the vehicle afterward.

Answers to the Test Questions for the Additional Test Questions Section 6

1. A	26. C	51. A	76. A
2. C	27. C	52. C	77. B
3. B	28. C	53. D	78. B
4. C	29. A	54. A	79. B
5. B	30. C	55. D	80. B
6. A	31. C	56. C	81. A
7. C	32. A	57. B	82. A
8. C	33. A	58. A	83. A
9. D	34. D	59. C	84. B
10. A	35. A	60. D	85. C
11. D	36. D	61. B	86. C
12. D	37. D	62. C	87. C
13. C	38. A	63. C	88. A
14. C	39. C	64. B	89. D
15. A	40. C	65. C	90. C
16. C	41. A	66. D	91. D
17. B	42. D	67. A	92. D
18. D	43. A	68. D	93. C
19. D	44. A	69. B	94. D
20. D	45. A	70. C	95. B
21. B	46. C	71. B	96. D
22. C	47. C	72. D	97. B
23. B	48. B	73. A	98. C
24. C	49. C	74. B	
25. C	50. A	75. C	

Explanations to the Answers for the Additional Test Questions Section 6

Question #1
Answer A is correct. Emission testing is not usually part of preliminary diagnosis.
Answer B is Wrong. Testing fuel pressure is part of a preliminary diagnosis for an SFI system.
Answer C is wrong. Checking for vacuum leaks is part of a preliminary diagnosis.
Answer D is wrong. A tamper inspection is part of a preliminary diagnosis.
Typical service procedures consist of verifying the complaint, doing a good visual inspection and simple test such as vacuum, compression, and cylinder leakage, retrieving DTCs, checking for any TSBs, looking at the scan tool data repairing the fault, and test-driving the vehicle afterward.

Question #2
Answer A is wrong. Technician B is also correct.
Answer B is Wrong. Technician A is also correct.
Answer C is correct. Both Technicians are correct. Low compression creates a misfire and results in more oxygen in the exhaust because of incomplete combustion. The oxygen sensor sees this as a lean exhaust condition. The computer will compensate by increasing long-term fuel trim commands.
Answer D is wrong. Both Technicians are correct.

Question #3
Answer A is wrong. The sensor works on a "wall vibration" principle, wherein a vibrating plate inside the sensor presses against a piezoelectric quartz crystal to generate an A/C voltage signal.
Answer B is correct. Checking to see if the RCM is retarding the ignition timing will verify the sensor is working.
Answer C is wrong. Testing for a shorted condition is premature for the same reason that answer A is premature.
Answer D is wrong. Testing sensor voltage also is premature but may be the logical next step.

Question #4
Answer A is wrong. A saturated charcoal canister is a possible cause for rich air/fuel mixtures off idle because of the purge cycle that occurs off idle at cruising speeds of 2500 rpm.
Answer B is Wrong. Air injection stuck in the upstream mode on a closed loop fuel control system will cause the computer to command a rich air/fuel mixture because of continuous air injection on the O_2 sensor.
Answer C is correct. A vacuum leak would not cause a rich air/fuel mixture at 2500 rpm because it is a larger percentage of intake air at idle than 2500 rpm.
Answer D is wrong. Fuel diluted engine oil will cause richer air/fuel mixtures because fuel vapors will be drawn into the intake manifold by the PCV system.

Question #5
Answer A is wrong. If the PCM were faulty, the vehicle would probably not run at all.
Answer B is correct. In limp-home mode, the PCM operates with a set of default values for fuel control. If the vehicle operates better in limp-home mode, a problem with air intake and fuel metering is indicated. Because the MAF sensor is linked to both, B is correct.
Answer C is wrong. O_2 sensor problems do not normally place a control system in a limp-home mode.
Answer D is wrong. If the catalytic converter is clogged, the vehicle will run poorly regardless of what mode the control system is in.

Question #6
Answer A is correct. This symbol represents a thermistor, which is the most common temperature sensor.
Answer B is Wrong. This symbol represents a potentiometer, such as a TPS or EGR valve position sensor.
Answer C is wrong. This symbol represents one kind of MAP sensor.
Answer D is wrong. This symbol represents an oxygen sensor.

Question #7
Answer A is wrong. Barometric pressure is a scan tool data item.
Answer B is Wrong. O_2 sensor voltage is a scan tool data item.
Answer C is correct. Fuel octane is not currently displayed on a scan tool data list.
Answer D is wrong. Miles per hour is a scan tool data item.

Question #8
Answer A is wrong. No sensor sends a variable resistance signal. There is no such thing, and the PCM cannot respond to changes in resistance.
Answer B is Wrong. No sensor sends a variable capacitance signal. There is no such thing, and the PCM cannot respond to changes in capacitance.
Answer C is correct. The knock sensor produces an oscillating analog voltage signal that can be checked with a DVOM or lab scope.
Answer D is wrong. The knock sensor does not produce a digital voltage signal.

Question #9
Answer A is wrong. CO_2 emissions are measured as a percentage of total exhaust gas.
Answer B is Wrong. O_2 emissions are measured as a percentage of total exhaust gas.
Answer C is wrong. CO emissions are measured as a percentage of total exhaust gas.
Answer D is correct. HC emissions are measured in parts per million (ppm).

Question #10
Answer A is correct. The presence of CO or HC in the air above the coolant in the radiator would indicate a combustion leak. By removing the radiator cap and holding the probe of a 4-gas analyzer just above the coolant with the engine running we test for HC in the radiator. Any sign of HC in the radiator is a sign of a leaking head gasket or head.
Answer B is Wrong. A burned exhaust valve would not contaminate the cooling system.
Answer C is wrong. A burned intake valve would not contaminate the cooling system.
Answer D is wrong. A vacuum leak would not contaminate the cooling system.

Question #11
Answer A is wrong. The catalytic converter can overheat when a spark plug wire is removed while the engine is running.
Answer B is Wrong. Damage to the ignition coil from high voltage carbon tracking can result when spark plug wires are removed while the engine is running.
Answer C is wrong. Spark voltage arcing to an ignition module inside the distributor cap can damage the module; this may occur when running an engine with a spark plug wire removed.
Answer D is correct. Damage to the ignition switch does not occur by removing a spark plug wire on a running engine.

Question #12
Answer A is wrong. If the fuel injector were leaking, CO would be higher.
Answer B is Wrong. A clogged air filter would cause a rich condition, and CO would be higher.
Answer C is wrong. Excessive fuel pressure would also produce a rich condition.
Answer D is correct. The vehicle has a lean misfire caused by a vacuum leak.

Question #13
Answer A is wrong. Measured engine speed may not be the speed commanded by the PCM.
Answer B is Wrong. Cranking rpm is engine speed during starting.
Answer C is correct. Based on inputs and calculations, this is the desired rpm.
Answer D is wrong. Throttle angle represents the throttle position expressed in degrees.

Question #14
Answer A is wrong. Technician A is correct in his method for checking for combustion chamber leaks.
Answer B is Wrong. Technician B is correct in his method for checking for vapor leaks.
Answer C is correct. Both Technicians are correct.
Answer D is wrong. Both Technicians are correct. Hydrocarbons are nothing more than unburned gas. A head gasket leaking or an EVAP system leak is releasing hydrocarbons. With the use of a gas analyzer leaking gas vapors can be found weather coming from the head gasket or any other part of a vehicle.

Question #15
Answer A is correct. Only Technician A is correct. When the air/fuel mixture is richer than 14.7 to1, the CO level will rise steadily in an almost linear fashion. This makes CO an excellent rich mixture indicator when using a four- or five gas analyzer.
Answer B is Wrong. The CO level is very low at 14.7 to 1 air/fuel ratios and remains almost flat as the mixture becomes leaner. Oxygen levels will rise rapidly as the mixture becomes leaner than 14.7 to 1 so this makes O_2 a good lean mixture indicator when measuring exhaust with a 4 or five-gas analyzer.
Answer C is wrong. Only Technician A is correct.
Answer D is wrong. Only Technician A is correct.IIC – unburned fuel CO = incomplete combustion and is a good running rich indicator CO_2 = engine efficiency complete combustion indicator O_2 = unused oxygen in the exhaust and is a good lean running indicator. The EVAP system is to control the release of HCs to the atmosphere.

Question #16
Answer A is wrong. Technician A is right that anything that causes uneven combustion or unequal compression can cause an engine to run rough.
Answer B is Wrong. Technician B is right that anything that causes uneven combustion can cause an engine to run rough.
Answer C is correct. Both Technicians are correct. These conditions also cause power loss, poor fuel economy, and increased emissions.
Answer D is wrong. Both Technicians are correct.

Question #17
Answer A is wrong. The oxygen sensor has no effect on cold starting. The 0.00-volt reading is normal for the composite vehicle with the key on, engine off.
Answer B is correct. The 0.35-volt reading indicates a fully warmed up engine, not the cold-start condition described. This high-temperature reading also supports the symptom described. This would cause a leaner mixture, rather than the richer mixture a cold engine requires.
Answer C is wrong. The IAT sensor is returning a correct signal for cold-start conditions. The IAT value should be relatively close to the coolant temperature value after sitting overnight.
Answer D is wrong. Given the data present, it would be premature to check PCM power and grounds. Sensor readings other than the ECT signal are within normal ranges.

Question #18
Answer A is wrong. Worn spark plugs would misfire more under load.
Answer B is Wrong. Dirty fuel injectors would misfire all the time.
Answer C is wrong. A defective knock sensor would not cause this problem.
Answer D is correct. If the TPS signal drops off to zero or very low voltage as the throttle moves, the PCM will think that the vehicle is decelerating rapidly and will cut off or reduce fuel. This, in turn, can cause a misfire and a stumbling or surging condition.
As a vehicle gets a few miles on it, the TPS may get a wear spot on the potentiometer. The wear spot will be the area on the TPS where it stays the most. For example if a person used the interstate to commute to work and they drove 70 MPH every day five days a week, 52 weeks a year, eventually the TPS might get a wear spot on it, right were it rides at 70 MPH. This would cause a vehicle to cut out and misfire at 70 MPH.

Question #19

Answer A is wrong. Ignition misfire would lead to high HC emissions.

Answer B is Wrong. Low cylinder compression would lead to high HC emissions.

Answer C is wrong. A lean mixture would lead to a misfire and high HC emissions.

Answer D is correct. A leaking exhaust system would not cause a misfire or high HC.

Question #20

Answer A is wrong. Excessive carbon deposits on the pistons could raise the compression ratio and increase combustion temperatures. This, in turn, could lead to high NOx emissions.

Answer B is Wrong. EGR is a principal NOx control method. A plugged EGR port would likely cause high NOx emissions.

Answer C is wrong. Hot spots in the combustion chamber could cause localized high combustion temperatures and increased NOx emissions.

Answer D is correct. A rich air/fuel mixture would cause high CO emissions, while NOx formation probably would be in the normal range. NOx is formed under high temperature conditions.

Question #21

Answer A is wrong. Checking grounds at this point would be premature. It is a voltage supply diagnosis.

Answer B is correct. Only Technician B is correct. A voltage drop of 0.6 volt between the battery and the PCM is excessive. The logical next steps would be to perform a voltage drop test and other resistance tests between the battery and the PCM to pinpoint the high resistance.

Answer C is wrong. Only Technician B is correct.

Answer C is wrong. Only Technician B is correct.

Question #22

Answer A is wrong. The PCM receives and processes sensor input information.

Answer B is Wrong. IAT data displayed on a scan tool can indicate both open-loop and closed-loop operation.

Answer C is correct. A scan tool displays input signals as interpreted by the PCM, and sent as serial data.

Answer D is wrong. Faults in circuit wiring can cause a different signal to be received by the PCM than the signal sent by the sensor.

Question #23

Answer A is wrong. Fuel pressure and volume could affect a hot re-start.

Answer B is correct. Because the engine cranks normally, battery capacity should be adequate and should not cause this hot re-start problem.

Answer C is wrong. The ignition coil could have an open winding when hot.

Answer D is wrong. A failing ignition pick-up may only act up when warm and could prevent the engine from starting until it cooled off.

Question #24

Answer A is wrong. 3.59 volts indicates a temperature of 0°C.

Answer B is Wrong. 2.27 volts would indicate a temperature of 40°C.

Answer C is correct. 0.46 volts indicates a temperature of 100°C.

Answer D is wrong. 0.84 volts would indicate a temperature of 80°C.

Question #25

Answer A is wrong. Technician B is also correct.

Answer B is Wrong. Technician A is also correct.

Answer C is correct. Both Technicians are correct. The primary coil windings share a common power feed, but have individual ground circuits. The secondary windings deliver spark simultaneously to both spark plugs in a series circuit. The schematic is illustrating a waste spark DIS system. On a waste spark system while one plug is firing on its compression stroke, its companion cylinder is firing on its exhaust stroke.

Answer D is wrong. Both Technicians are correct.

Question #26
Answer A is wrong. Relays are a part of the output control circuit.
Answer B is Wrong. Solenoids are a part of the output control circuit.
Answer C is correct. Sensors are input signals. There is no need for a driver-controlled input.
Answer D is wrong. Display devices are a part of the output control circuit.

Question #27
Answer A is wrong. This amount exceeds the 1.5 limit.
Answer B is Wrong. The limit is not measured in percent.
Answer C is correct. The OBD II system is designed to turn the MIL on when it determines the emission levels exceed 1.5 times the allowable standard.
Answer D is wrong. The limit is not measured in percent.
The OBD II systems must illuminate the malfunction indicator lamp (MIL) if the vehicle conditions would allow emissions to exceed 1.5 times the allowable standard for that model year based on a Federal Test Procedure (FTP). When a component or strategy failure allows emissions to exceed this level, the MIL is illuminated to inform the driver of a problem and a diagnostic trouble code is stored in the power train control module (PCM).

Question #28
Answer A is wrong. Adjusting the idle speed would not control a rich condition.
Answer B is Wrong. Increasing fuel delivery is the opposite action to correction a rich condition.
Answer C is correct. Decreasing fuel delivery would compensate for a rich condition.
Answer D is wrong. Energizing the purge solenoid would only richen the mixture.

Question #29
Answer A is correct. EGR valve failure usually causes a lack of EGR. This will not affect HC emissions.
Answer B is Wrong. Cylinder misfire can result in elevated HC readings.
Answer C is wrong. Incomplete combustion can cause high HC readings.
Answer D is wrong. An excessively lean air/fuel mixture can result in high HC readings, due to a lean misfire.

Question #30
Answer A is wrong. A leaking fuel pump check valve would cause fuel to drain back to the gas tank and increase cranking time needed to build fuel pressure.
Answer B is Wrong. A coolant sensor reading that is too high for a cold engine would make the computer supply a fuel mixture that is too lean for easy starting.
Answer C is correct. While a bad PCM is possible, it is the LEAST likely to cause this condition. Additionally, a failed PCM would probably create other problems as well a hard start condition.
Answer D is wrong. Many computers receive a CRANK input to increase fuel delivery during starting. A problem in this circuit could cause hard starting.

Question #31
Answer A is wrong. Technician A is correct that TPS and most MAP sensors are resistive devices that operate on a 5-volt reference voltage from the PCM.
Answer B is Wrong. Technician B is correct that TPS and analog MAP sensors both provide low-voltage at idle and almost reference voltage at wide-open throttle.
Answer C is correct. Both Technicians are correct.
Answer D is wrong. Both Technicians are correct.

Question #32
Answer A is correct. Sensors cannot have any values changed by a scan tool.
Answer B is Wrong. A scan tool enables a technician to read trouble codes.
Answer C is wrong. A scan tool allows a comparison of KOEO and KOER data.
Answer D is wrong. A scan tool allows a technician to examine sensor data.

Question #33

Answer A is correct. The conditions do not point to a faulty ECM. The computer is recognizing a rich condition and sending a lean command to the injectors.
Answer B is Wrong. Leaking fuel injectors could cause an excessive rich condition as the question states.
Answer C is wrong. High fuel pressure could cause an excessive rich condition as the question states.
Answer D is wrong. The EVAP system purging at idle could cause overly rich mixtures as the question states.

Question #34

Answer A is wrong. A jumped timing belt would cause other drivability problems.
Answer B is Wrong. Retarded ignition timing would tend to slow the idle rather than speed it up. In this case, the composite vehicle has Distributorless Ignition (EI). Ignition timing is not adjustable.
Answer C is wrong. Fuel pressure that is too high would not cause a high idle.
Answer D is correct. An intake manifold vacuum leak would allow unmetered air into the intake and raise the idle. The PCM would try to close the TAC to compensate, which is why the TAC percentage is zero.

Question #35

Answer A is correct. Only Technician A is correct. Scan tool power is supplied to terminal #16 on the composite vehicle. A shorted wire would blow fuse 74 and prevent the scan tool from powering up.
Answer B is Wrong. The OBD II system on the composite vehicle will illuminate the MIL for emissions related failures only; this condition does not affect emissions.
Answer C is wrong. Only Technician A is correct.
Answer D is wrong. Only Technician A is correct.

Question #36

Answer A is wrong. A diode tester does not measure either system voltage or signal interference.
Answer B is Wrong. An ohmmeter measures resistance, not system voltage or signal interference.
Answer C is wrong. A voltmeter does not sample and display voltage fast enough to detect signal interference.
Answer D is correct. An oscilloscope can measure and display both system voltage and high-frequency signal interference caused by RFI.

Question #37

Answer A is wrong. Emission routing information is not found in the owner's guide.
Answer B is Wrong. Emission routing information is not found in the trunk area.
Answer C is wrong. Emission routing information is not found in the audio guide.
Answer D is correct. The emission label is specific to that particular vehicle, lock-up and area (high altitude, etc.). The emission label is located under the hood in areas such as the hood, valve cover, fan shroud or inner fender.

Question #38

Answer A is correct. Only Technician A is correct. On many fuel-injection systems, high voltage from the TPS can put the vehicle in a clear-flood mode and disable the fuel injectors while cranking the engine.
Answer B is Wrong. The injectors would operate electrically without the fuel pump working, but the injectors are not working.
Answer C is wrong. Only Technician A is correct.
Answer D is wrong. Only Technician A is correct.

Question #39

Answer A is wrong. Technician A is right that a shorted actuator can cause excessive current, which can damage the output driver transistors in a computer. The computer is subject to repeated failure until the defective actuator is replaced.
Answer B is Wrong. Technician B is right that a faulty voltage suppression, or clamping, diode can let an inductive voltage spike drive excessive current through an actuator circuit, which can damage the output driver transistor in a computer. The computer is subject to repeated failure until the defective diode is replaced.
Answer C is correct. Both Technicians are correct.
Answer D is wrong. Both Technicians are correct.

Question #40

Answer A is wrong. Technician A is correct. High HCs in the converter combining with O_2 can significantly raise the converter temperature, thus resulting in damage.

Answer B is Wrong. Technician B is correct. The catalytic converter will successfully reduce CO and HC levels and further convert them to CO_2 and H_2O, providing proper combustion occurs in the engine.

Answer C is correct. Both Technicians are correct.

Answer D is wrong. Both Technicians are correct.

Question #41

Answer A is correct. This statement refers to the ratio of air related to fuel by weight, required to support proper and complete combustion at sea level. A typical zircronia O_2 sensor sends a low voltage signal (under 450MV) when it sees a high level of oxygen in the exhaust. It sends a high voltage signal (over 450MV up to 1 volt) when it sees a low level of oxygen in the exhaust.

Answer B is Wrong. This mixture would be richer than 14.7:1.

Answer C is wrong. This mixture would be slightly leaner than 14.7:1.

Answer D is wrong. This mixture would vary according to the ambient conditions.

Question #42

Answer A is wrong. Excessive engine oil pressure can pump up camshaft followers and cause valves to remain open which will cause a misfire.

Answer B is Wrong. A broken rocker arm shaft pedestal will create excessive valve lash and can cause engine misfire.

Answer C is wrong. A leaking fuel-pressure regulator diaphragm will allow raw fuel into the intake manifold and create a rich misfire condition. Cylinder misfire faults are generated by an abnormal speed-up of the rpm reference signal. Mechanical faults and ignition faults are typically the source of misfire faults.

Answer D is correct. Low alternator output would have no effect on engine misfire.

Question #43

Answer A is correct. Only Technician A is correct. Excessive crankcase blow-by will cause excessive build-up of carbon deposits around the throttle plate of a throttle body.

Answer B is Wrong. In many cases the result of a dirty throttle body is a vehicle that stalls at start up and when coming to a stop.

Answer C is wrong. Only Technician A is correct.

Answer D is wrong. Only Technician A is correct.

Question #44

Answer A is correct. Only Technician A is correct. Always check circuit wiring before condemning a sensor.

Answer B is Wrong. The ECT and any other sensor should be tested as it is installed and operating as part of the system. Removing a temperature sensor and trying to test it "in hot water with an ohmmeter" would be an enormous waste of time.

Answer C is wrong. Only Technician A is correct.

Answer D is wrong. Only Technician A is correct.

Question #45

Answer A is correct. Because the air/fuel mixture is enriched during acceleration, a vacuum leak is LEAST likely to cause a loss of power under this condition. A vacuum leak typically does not cause a problem off idle. Most vacuum leaks cause ether a high idle or if the leak is large enough a rough idle or rough running engine that is not felt off idle.

Answer B is Wrong. A clogged fuel filter could cause a loss of power due to reduced fuel delivery.

Answer C is wrong. An overtorqued knock sensor could excessively retard timing.

Answer D is wrong. Detonation caused by poor fuel quality could cause the PCM to retard timing in response to knock sensor signals, if so equipped.

Question #46

Answer A is wrong. Technician B is also correct.

Answer B is Wrong. Technician A is also correct.

Answer C is correct. Both Technicians are correct. A mass airflow performance code could be set by a problem with the sensor itself or from a mechanical condition causing a problem with airflow through the engine.

Answer D is wrong. Both Technicians are correct. Anytime a PCM stores a fault there is a good possibility that the component itself is at fault. A component should never be replaced without testing the component and circuit according to the manufacturer's test procedure. Sometimes a fault is set due to malfunctioning systems that are not directly connected with the failed circuit. Always test the faulted circuit according to manufacturer's test procedures.

Question #47

Answer A is wrong. In most fuel-injection systems, a shorted injector driver would keep the injector energized continuously and cause an overly rich air/fuel mixture at all engine speeds. This does not match the symptoms.

Answer B is Wrong. A worn intake cam lobe would reduce airflow into the cylinder and O_2 levels would be low, not high as shown at idle. The engine would also misfire at high speeds.

Answer C is correct. A vacuum leak will cause a lean misfire in a cylinder at idle but the condition will disappear when the throttle is opened. The high oxygen readings at idle also indicates this as the problem.

Answer D is wrong. A fouled spark plug would affect HC emissions at idle and 2000 rpm. The misfire would also be noticeable at both engine speeds.

Question #48

Answer A is wrong. The ECT signal can cause a rich mixture.

Answer B is correct. The O_2 sensor is working, showing a rich exhaust condition not causing it,

Answer C is wrong. The TPS can cause enrichment.

Answer D is wrong. An incorrect MAP signal can cause a rich mixture.

Question #49

Answer A is wrong. Technician A is right that overall low block learn (long-term fuel trim) indicates an overall rich condition, but that is only part of the answer.

Answer B is Wrong. Technician B is right that high low block learn (long-term fuel trim) at idle can indicate a vacuum leak because the leak is a larger percentage of intake air at idle, but that is only part of the answer.

Answer C is correct. Both Technicians are correct.

Answer D is wrong. Most adaptive strategies have two parts: Short Term Fuel Trim (STFT) and Long Term Fuel Trim (LTFT). Short-term strategies are those immediately enacted by the computer to overcome a change in operation. These changes are temporary. Long-term strategies are based on the feedback about the short-term strategies.

Question #50

Answer A is correct. Only Technician A is correct. Technician A has identified one of the reasons that an ignition coil must have some voltage in reserve for high-load conditions.

Answer B is Wrong. During normal operating conditions at idle or cruise, the ignition will need to output usually less than 25 percent of its capability. The reserve is used for various load and temperature conditions.

Answer C is wrong. Only Technician A is correct.

Answer D is wrong. Only Technician A is correct. When dealing with secondary voltage, there are three types of voltage to deal with: available voltage (maximum coil output), required voltage (voltage required to jump the plug gap), and reserve voltage (voltage left over).

Question #51

Answer A is correct. A wide spark plug gap would increase firing KV and shorten spark duration and the exact opposite is seen on #1 cylinder.

Answer B is Wrong. A fouled spark plug would show low firing KV and longer spark duration.

Answer C is wrong. Low cylinder compression could cause the waveform problem on #1 cylinder.

Answer D is wrong. A bridged spark plug gap would cause the waveform problem seen on #1 cylinder.

The picture is that of secondary ignition. Cylinder #1 has a low firing line and a wide spark line. The spark line voltage is also low. Activity in the combustion chamber that would require more voltage would be things such as high compression, a lean mixture, worn spark plugs and wide spark plug gags. A fouled spark plug, low cylinder compression or bridged gap all would require less secondary voltage as seen.

Question #52

Answer A is wrong. Technician B is also correct.

Answer B is Wrong. Technician A is also correct.

Answer C is correct. Both Technicians are correct. A faulty fuel-pressure regulator can leak allowing fuel into the intake manifold increasing CO emissions. A rich-running engine lowers combustion chamber temperature that reduces the formation of NO_X. By replacing the fuel pressure regulator the combustion chamber temperature will return to normal operating temperature reveling high NO_X emissions if other systems are at fault.

Answer D is wrong. Both Technicians are correct.

Question #53

Answer A is wrong. Technician B is also correct.

Answer B is Wrong. Technician A is also correct.

Answer C is correct. Both Technicians are correct. In the sensor shown, terminal A would have VREF applied and terminal B would be grounded. The signal voltage would be low, generally less than 1 volt, and increase as the sensor's wiper arm rotated clockwise.

Answer D is wrong. Both Technicians are correct. The PCM typically sends a 5-volt reference to most sensors. The sensor is grounded through the PCM or remotely, and a signal return is sent back to the PCM for the PCM to read the signal voltage of the sensor. The TPS sends a low voltage at idle and a high voltage at WOT because of the wiper position.

Question #54

Answer A is correct. Only Technician A is correct. System polarity must be observed when testing devices incorporating spike suppression diodes in order to prevent diode damage.

Answer B is Wrong. While the solenoid will energize regardless of system polarity, the diode will blow open if polarity is reversed to the solenoid. Technician B's statement is wrong since damage will occur.

Answer C is wrong. Only Technician A is correct.

Answer D is wrong. Only Technician A is correct.

Question #55

Answer A is wrong. Technician A is wrong because secondary air injection has no effect on reducing NO_X emissions.

Answer B is Wrong. Technician B is wrong because secondary air is supplied to the converter after the engine is warmed up and in closed loop, not during warm-up.

Answer C is wrong. Neither Technician is correct. In a three-way catalytic converter the first section is designed to convert NO_X back to the oxygen and nitrogen, the AIR system is plumbed in after the first section so secondary air can be used in the conversion process of HC and CO.

Answer D is correct. Neither Technician is correct.

Question #56

Answer A is wrong. HC is a byproduct of combustion and would not cool combustion chamber temperatures.

Answer B is wrong. NOx is a byproduct of elevated combustion chamber temperatures.

Answer C is correct. Exhaust gas is introduced into the combustion chamber by the EGR valve to cool chamber temperatures. Exhaust gas is inert and does not support combustion. In the combustion chamber, via the EGR, this inert gas slows the combustion burn, thus reducing temperatures.

Answer D is wrong. Air introduced into the combustion chamber will not lower temperatures. HC = unburned fuel NOX is a combination of nitrogen and oxygen which is formed under high heat conditions.

Exhaust gas is what the EGR system uses to lower combustion chamber temperature. Air is required for the burn to take place in the combustion chamber.

Question #57

Answer A is wrong. Performing voltage tests on computer ground circuits is not the first step in diagnosing drivability concerns.

Answer B is correct. Only Technician B is correct. Reviewing vehicle service history and verifying the customer complaint are normally performed when starting a drivability diagnosis.

Answer C is wrong. Only Technician B is correct.

Answer D is wrong. Only Technician B is correct.

Question #58

Answer A is correct. Only Technician A is correct. Retarded ignition timing will reduce both power and gas mileage.

Answer B is Wrong. Power loss and poor fuel mileage are not symptoms of excessive timing advance.

Answer C is wrong. Only Technician A is correct.

Answer D is wrong. Only Technician A is correct. Retarded ignition timing will reduce both power and gas mileage, as well as cause long crank time before starting and engine over heating. Excessive timing advance will cause engine ping or spark knock and possibly a engine miss at cruising speed.

Question #59

Answer A is wrong. Technician B is also correct.

Answer B is Wrong. Technician A is also correct.

Answer C is correct. Both Technicians are correct. A fixed WOT signal from the throttle position sensor will cause the computer to command a rich air/fuel mixture and increase emissions levels.

Answer D is wrong. Both Technicians are correct.

Question #60

Answer A is wrong. The oxygen sensors are reading normally.

Answer B is Wrong. Low fuel pressure would lean out the engine and drive the oxygen sensors low. The oxygen sensors are reading normally.

Answer C is wrong. A faulty injector would affect the oxygen sensor readings and more than likely set a fault and illuminate the MIL.

Answer D is correct. The scan tool data shows the TCC solenoid going on, then off. This is not normal operation of the TCC solenoid. A malfunction in this circuit can feel like an engine miss.

Question #61

Answer A is wrong. A clogged air filter will raise CO emissions and have little effect on HC emissions.

Answer B is correct. Only Technician B is correct. Overly advanced ignition timing can keep combustion temperatures from reaching their peak because combustion occurs before cylinder pressure reaches its peak. The resulting lower combustion temperature quenches the mixture and leaves unburned hydrocarbon (HC).

Answer C is wrong. Only Technician B is correct.

Answer D is wrong. Only Technician B is correct.

Question #62

Answer A is wrong. Lower CO emissions do not result from lower combustion chamber temperatures.

Answer B is Wrong. Lower CO_2 emissions do not result from lower combustion chamber temperatures.

Answer C is correct. NO_x is formed at high temperatures. Excessive lean mixtures can increase temperatures.

Answer D is wrong. High O_2 emissions do not result from lower combustion chamber temperatures.

Question #63

Answer A is wrong. The trace is not normal.

Answer B is Wrong. The technician performed the test correctly.

Answer C is correct. The dip in the trace is an open or high-resistance condition in the TPS. It must be replaced.

Answer D is wrong. Wide-open throttle is at the end of the trace. This doesn't apply.

Question #64

Answer A is wrong. A balance test is used to locate injector flow rate differences. This test should not be performed until fuel pressure has been tested to make sure fuel pressure is correct and that pressure does not leak down.

Answer B is correct. Only Technician B is correct. High fuel pressure is a likely cause of high CO emissions and a failed IM-test.

Answer C is wrong. Only Technician B is correct.

Answer D is wrong. Only Technician B is correct.

Question #65

Answer A is wrong. Hard starting and engine stalling when warm can result from fuel enrichment caused by an open ECT sensor circuit that indicates subzero temperature.

Answer B is Wrong. The low-temperature signal sent by an open ECT sensor can cause the PCM to inhibit torque converter lock-up.

Answer C is correct. A lean air/fuel mixture cannot result from the low-temperature signal sent by an open ECT sensor circuit. Rather, the mixture will be overly rich. Answer A is wrong. Hard starting and engine stalling when warm can result from fuel enrichment caused by an open ECT sensor circuit that indicates subzero temperature.

Answer D is wrong. EGR operation will be altered or cut off entirely at low temperature.

An ECT is a negative temperature coefficient sensor meaning as temperature goes up resistance goes down. An open at terminal 56 would make the PCM see a cold engine coolant temperature regardless of actual engine temperature.

Question #66

Answer A is wrong. The vehicle speed has to be ABOVE a certain threshold for operation.

Answer B is Wrong. Coolant temperature has to be ABOVE a certain threshold for operation.

Answer C is wrong. Neither Technician is correct. Purge of the EVAP system typically happens at speeds above Idle rpm and on a warm engine. Purging of the EVAP at Idle or on a cold engine will cause drivability problems at idle, on acceleration and during engine warm-up.

Answer D is correct. Neither Technician is correct.

Question #67

Answer A is correct. The composite vehicle has a returnless fuel system. There is no vacuum hose to the regulator. The regulator is in the tank.

Answer B is Wrong. This would cause low fuel pressure and volume.

Answer C is wrong. This would reduce the flow of fuel, thus reducing pressure as well as volume.

Answer D is wrong. Excessive voltage drop on the ground side of the fuel pump will also cause low pressure and volume.

Question #68

Answer A is wrong. If the EGR valve stays open at wide-open throttle or other heavy-load operation, it will not have much effect on manifold vacuum, but it can reduce power by diluting the air/fuel mixture.

Answer B is Wrong. Retarded camshaft timing will reduce both intake manifold vacuum and engine power.

Answer C is wrong. A restricted exhaust will reduce power under load, as well as under other operating conditions. It also will reduce manifold vacuum because of excessive back pressure.

Answer D is correct. A rich air/fuel ratio is LEAST likely to reduce power, in fact it is supplied to increase power, and it will not noticeably affect manifold vacuum.

Question #69

Answer A is wrong. Engine surging can be a symptom of low fuel pressure.

Answer B is correct. Black exhaust smoke usually indicates a rich air/fuel mixture, which would not result from low fuel pressure.

Answer C is wrong. Lack of power can be a symptom of low fuel pressure.

Answer D is wrong. Acceleration stumble can be a symptom of low fuel pressure.

Black smoke = unburned fuel. Blue or grayish smoke = oil. White smoke = coolant.

Question #70

Answer A is wrong. Technician B is also correct.

Answer B is wrong. Technician A is also correct.

Answer C is correct. Both Technicians are correct.

Answer D is wrong. Both Technicians are correct.

Question #71

Answer A is wrong. This schematic is for a sequential (SFI) fuel-injection system.

Answer B is correct. Only Technician B is correct. This schematic shows individual ground circuits for each injector in this sequential system.

Answer C is wrong. Only Technician B is correct.

Answer D is wrong. Only Technician B is correct.

Question #72

Answer A is wrong. Answers B and C are also correct. Restricted injectors would cause the vehicle to starve for fuel.

Answer B is Wrong. Answers A and C are also correct. A vacuum leak would lean the engine.

Answer C is wrong. Answers A and B are also correct. Intake valve deposits act like sponges soaking up the fuel.

Answer D is correct. Any of the above can disrupt fuel delivery and cause a variety of drivability symptoms.

Question #73

Answer A is correct. A thermo-time switch only operates during cold startup when NO_x emissions are not formed. Additionally, an I/M test would not be performed on a cold engine.

Answer B is Wrong. An inoperative EGR valve will increase NO_x emissions.

Answer C is wrong. A defective catalytic converter can increase NO_x emissions.

Answer D is wrong. A catalytic converter that has not heated up will have poor efficiency and can increase NO_x emissions.

Question #74

Answer A is wrong. A restricted PCV valve is unlikely to increase CO emissions until crankcase pressure becomes so high that vapors are forced back through the clean-air hose to the intake airflow. This probably would not occur during the short time of an emissions test.

Answer B is correct. Only Technician B is correct. The vapor storage canister is normally purged at cruising speed above 20 mph. A fuel-saturated canister would result in increase the purge rate with subsequently higher CO emissions.

Answer C is wrong. Only Technician B is correct.

Answer D is wrong. Only Technician B is correct.

Question #75

Answer A is wrong. Technician A is correct that coolant can enter the combustion chamber from a leaking head gasket and contaminate the oxygen sensor as it passes out in the exhaust.

Answer B is Wrong. Technician B is correct that ethylene glycol in the coolant, as well as silicate additives, can contaminate the oxygen sensor.

Answer C is correct. Both Technicians are correct.

Answer D is wrong. Both Technicians are correct.

Question #76

Answer A is correct. Only Technician A is correct. The knock sensor is checked using a small hammer and timing light or scan tool.

Answer B is Wrong. Timing is non-adjustable in the composite vehicle.

Answer C is wrong. Only Technician A is correct.

Answer D is Wrong. Only Technician A is correct.

Question #77

Answer A is wrong. If the transmission pressure control solenoid wire was shorted to ground (PCM terminal50), it would cause the mainline pressure to be low. The result would be soft, early shifts not harsh, late shifts.

Answer B is correct. Only Technician B is correct. The purpose of the loaded-mode emissions test is to sample emissions during simulated real world driving on a dynamometer. Any power train problems may affect how accurate the readings will be. Late transmission shifting can increase engine rpm throughout the test and raise overall emissions output.

Answer C is wrong. Only Technician B is correct.

Answer D is wrong. Only Technician B is correct.

Question #78

Answer A is wrong. Condensation occurs when the atomized fuel separates from the air. This does not happen when injected fuel hits carbon deposits on the back of the intake valve.

Answer B is correct. Carbon deposits on the backs of intake valves can absorb part of the injected fuel. Carbon buildup on the intake valve can cause a vehicle to stumble due to a lean air fuel mixture. The reason for the lean air fuel mixture is fuel absorption because the carbon acts like a sponge.

Answer C is wrong. Vaporization is the process of turning fuel into a gaseous state. It has nothing to do with problems caused by intake valve deposits.

Answer D is wrong. Puddling is the result of condensation in the intake manifold of a carbureted engine and would be an irrelevant answer to this question.

Question #79

Answer A is wrong. Service bulletins provide useful information when diagnosing all types of problems.

Answer B is correct. The question suggests a non-computer-related problem. Any mechanical failure must be addressed prior to scan testing.

Answer C is wrong. The vehicle service manual often is necessary for proper diagnosis.

Answer D is wrong. A vacuum gauge is very useful for powertrain mechanical failure diagnosis.

Question #80

Answer A is wrong. Spark plug overheating can result from incorrect ignition timing.

Answer B is correct. These fused deposits usually result from intermittent overheating that can be caused by hard acceleration or heavy engine loads.

Answer C is wrong. Pre-ignition can result form incorrect or advanced ignition timing.

Answer D is wrong. Carbon fouling can result form an ignition misfire.

Question #81

Answer A is correct. The gas readings show normal operation and converter action.

Answer B is Wrong. If the vehicle had a vacuum leak, the idle HC and O_2 readings would be elevated.

Answer C is wrong. These readings indicate a properly functioning catalytic converter that is using oxygen for the conversion process.

Answer D is wrong. These readings indicate a properly functioning catalytic converter that is using oxygen for the conversion process. Typical 4-gas readings of a good-running vehicle are HC = 50 PPM or less, CO = .2 to .5%, O_2 = 0 to 2%, CO_2 = 13% or higher HC, CO and O_2 should lower at higher rpm and CO_2 should increase at higher rpm.

Question #82

Answer A is correct. Only Technician A is correct.

Answer B is wrong. Only Technician A is correct.

Answer C is wrong. Only Technician A is correct.

Answer D is wrong. Only Technician A is correct.

Question #83

Answer A is correct. Only Technician A is correct. Most multiport fuel injectors are mounted in the intake manifold and are susceptible to increased heat transfer from the engine.

Answer B is Wrong. TBI injectors are not subjected to the high temperatures that MFI and SFI injectors are, and therefore, are not as susceptible to tip deposits.

Answer C is wrong. Only Technician A is correct.

Answer D is wrong. Only Technician A is correct.

Question #84

Answer A is wrong. High float level on a carbureted engine would cause a rich mixture and high CO readings. HC would be closer to normal.

Answer B is correct. A large vacuum leak will cause a lean misfire and resulting high HC emissions. The O_2 sensor voltage is low because of the excess oxygen passing through the combustion chambers.

Answer C is wrong. A dirty air filter may restrict intake airflow and cause high CO emissions.

Answer D is wrong. High fuel pressure can increase fuel delivery and cause a rich air/fuel mixture with related high CO emissions.

Question #85

Answer A is wrong. Technician A is correct. Refer to Ohm's law: resistance and current flow are inversely proportionate when voltage remains the same. Too much current flow due to a low resistance problem can damage other electrical components or circuits.

Answer B is Wrong. Technician B is correct. High resistance will add another load in the circuit, thus not leaving proper current flow to operate components or circuits.

Answer C is correct. Both Technicians are correct.

Answer D is wrong. Both Technicians are correct.

Question #86

Answer A is wrong. A failed block ground would cause other problems before shutting down the cold-start injector.

Answer B is Wrong. The oxygen sensor would not cause this problem.

Answer C is correct. The thermo-time switch controls the cold-start injector operation. The cold start injector is an auxiliary injector operated by the starter solenoid when temperatures are cold. The thermo-time switch is what controls this action.

Answer D is wrong. The ECT does not control cold-start injector operation.

Question #87
Answer A is wrong. An injector balance test should only be performed after testing the fuel pressure.
Answer B is Wrong. Injector flow testing should only be performed after testing the fuel pressure.
Answer C is correct. A fuel pressure test should be done first along with fuel volume. Lean surge and hard starting are both symptoms of inadequate fuel supply. The first test for inadequate fuel supply would be pressure and volume.
Answer D is wrong. Injector sound testing only confirms operation.

Question #88
Answer A is correct. A leaking fuel injector could cause a rich fuel trim on the bank the injector is on.
Answer B is Wrong. An intake manifold leak would not cause a rich mixture.
Answer C is wrong. An EGR valve stuck closed would not cause a rich mixture.
Answer D is wrong. Low charging system voltage would not cause a rich mixture. A pinched fuel return line would cause high fuel pressure, which would result in a rich air/fuel mixture. A rich mixture has a lack of oxygen present in it which would cause the O_2 sensor to give a continuously high voltage signal (1 volt) this would cause the fuel trim rich fault after the PCM attempted to lean the air/fuel mixture and failed. The difference from bank 1 and bank 2 is that bank 1 refers to the side of the engine where the number 1 cylinder is usually located.

Question #89
Answer A is wrong. The use of silicone grease in this location is unrelated to engine heat.
Answer B is Wrong. The module casing must be grounded as part of the circuit on most types of systems. The properties of the actual material that houses the module provides protection against moisture and other elements.
Answer C is wrong. Module ground is established through the electrical connector.
Answer D is correct. Silicone grease under the module helps to dissipate heat from the module.

Question #90
Answer A is wrong. Technician A is correct that linear EGR valve operation is usually displayed as an item on the data stream accessible with a scan tool.
Answer B is Wrong. Technician B is correct that a positive back pressure EGR valve can be tested off the car by applying air pressure to the exhaust inlet and then applying vacuum to the vacuum diaphragm. The valve should open.
Answer C is correct. Both Technicians are correct.
Answer D is wrong. Both Technicians are correct.

Question #91
Answer A is wrong. A misfire would increase HO and O_2.
Answer B is Wrong. Low fuel pressure may show as a lean condition.
Answer C is wrong. A vacuum leak would show an increase of O_2 and HC.
Answer D is correct. The readings at an idle are very good. At 2000 rpm, it became over rich as indicated by very high CO and HC. There is a fuel delivery control problem or purge problem with this vehicle. Typical 4-gas readings of a good running vehicle are HC = 50 PPM or less, CO = .2 to .5%, O_2 = 0 to 2%, CO_2 = 13% or higher HC, CO and O_2 should lower at higher rpm and CO_2 should increase at higher rpm.

Question #92
Answer A is wrong. EVAP system purging at idle results in higher than normal CO readings due to excess fuel.
Answer B is Wrong. Lower than specified fuel pressure results will not cause high CO readings.
Answer C is wrong. Neither Technician is correct. EVAP purging at idle causes a rich condition that causes CO to increase, not decrease. Low fuel pressure would lean the system causing CO to lower; it could also lean to the point of lean misfire, which would increase HC.
Answer D is correct. Neither Technician is correct.

Question #93

Answer A is wrong. Higher than normal combustion chamber temperature sod not result in elevated CO readings.
Answer B is Wrong. A lean air/fuel mixture most often results in a lean misfire and higher HC emissions.
Answer C is correct. When a fuel injector fails to close completely, the resulting drip of fuel will cause a rich air/fuel mixture and subsequent increase in CO emissions.
Answer D is wrong. Cylinder misfiring may result in lower, not higher, CO emissions. HC = unburned fuel. CO= incomplete combustion and is a good-running rich indicator. CO_2 = engine efficiency complete combustion indicator. O_2 = unused oxygen in the exhaust and is a good lean-running indicator.

Question #94

Answer A is wrong. A stuck-closed PCV valve will increase crankcase pressure and can cause oil to be pushed past gaskets.
Answer B is Wrong. A stuck-closed or blocked PCV valve reduces airflow into the engine and can lower idle speed.
Answer C is wrong. A stuck-closed or blocked PCV valve will increase crankcase pressure and cause oil to be forced through the breather hose into the air cleaner housing.
Answer D is correct. A stuck-closed or blocked PCV valve will lower idle speed because there is less air flowing into the engine. The computer will try to compensate for low idle speed by increasing idle air control opening.

Question #95

Answer A is wrong. Technician A is wrong because purging the charcoal canister will enrich the air/fuel mixture and increase CO emissions, yet the CO levels were normal.
Answer B is correct. Only Technician B is correct. The I/M240 analyzer collects vehicle exhaust with a constant volume sampler or CVS. Any external fuel leaks can allow HC to be drawn into the sampling hose and cause a test failure showing high HC levels throughout the drive trace that do not coincide with engine operation.
Answer C is wrong. Only Technician B is correct.
Answer D is wrong. Only Technician B is correct.

Question #96

Answer A is wrong. A faulty power steering switch would not cause this condition.
Answer B is Wrong. The battery voltage is within specifications.
Answer C is wrong. The crankshaft position sensor is sending a reference rpm.
Answer D is correct. The composite vehicle uses an immobilizer anti-theft system. In short, if an attempt is made to start the vehicle with an invalid ignition key, the ECM will disable the fuel injectors to kill the engine.

Question #97

Answer A is wrong. A vacuum leak usually affects long-term fuel trim more at idle than at higher rpm because an intake leak adds a greater percentage of air to the air/fuel mixture at idle.
Answer B is correct. Only Technician B is correct. A clogged fuel filter or weak fuel pump would affect fuel delivery equally at all engine speeds, and long-term fuel trim would more likely be equally high at all speeds to compensate.
Answer C is wrong. Only Technician B is correct.
Answer D is wrong. Only Technician B is correct.

Question #98

Answer A is wrong. Both Technicians are correct.
Answer B is Wrong. Both Technicians are correct.
Answer C is correct. Both Technicians are correct. Incorrect shift points may be the result of the transmission controller (function) or the transmission itself. This can cause perceived engine performance problems and affect emissions. Erratic overdrive shifting or torque converter clutch problems can easily be misinterpreted as fuel or ignition related.
Answer D is wrong. Both Technicians are correct.
The transmission is the cause of many drivability complaints. While a drivability technician does not have to know how to rebuild an automotive transmission, the technician does need to be able to distinguish a transmission fault from a drivability fault. The lockup clutch engaging too soon will cause a lack of power and a vibration. A lockup converter that does not disengage will cause stalling when coming to a stop. A transmission shifting in and out of overdrive will feel like an engine surge.

Glossary

The reference materials and questions for the L1 Test use electronic and emission terms and acronyms that are consistent with the industry-wide SAE standards J1930 and J2012. Some of these terms are listed below.

Calculated Load Valve The percentage of engine capacity being used, based on current airflow divided by maximum airflow.

Data Link Connector (DLC) The standardized plug that is used to connect the scan tool to the computer.

Diagnostic Trouble Codes(DTC) Codes stored by the computer when a problem is detected and read using a scan tool. Each code corresponds to a particular problem. When a DTC is referred to in an L1 test question, the number and description will both be given. For instance, P0114 = Intake Air Temperature Circuit Intermittent.

Distributor Ignition(DI) An ignition system that uses a distributor.

Electronic Ignition (EI) An ignition system that has coils dedicated to specific spark plugs and does not use a distributor; often referred to as distributorless ignition.

Freeze-Frame Operating conditions that are stored in the memory of the PCM at the instant a diagnostic trouble code is set.

Fuel Trim (FT) Fuel delivery adjustments based on closed-loop feedback. Valves above the central value (0%) indicate increased injector pulse width. Valves below the central value (0%) indicate decreased injector pulse width. Short Term Fuel Trim is based on rapidly switching oxygen sensor values. Long Term Fuel Trim is a learned value used to compensate for continual deviation of the Short Term Fuel Trim from its central value.

Generator J1930 term for alternator (generating device that uses a diode rectifier).

I/M Tests Inspection and Maintenance Tests; vehicle emissions tests required by state governments. Some common types of I/M tests include:

No-Load Tests that measure HC emissions in parts per million (ppm) and CO emissions in percent, while the vehicle is in neutral. Examples are idle and two-speed.

Acceleration Simulation Mode (ASM) Loaded-mode steady-state tests that measure HC, CO and NO_x emissions while the vehicle is driven on a dynamometer at a fixed speed and load. ASM5015 is a test at 15 mph with a load equivalent to 50% of the power needed to accelerate the vehicle at 3.3 mph per second. ASM2525 is a test at 25 mph with a load of 25% of the same power.

I/M240 A loaded-mode transient test that measures HC, CO, CO_2, and NO_x emissions in grams/mile second by second, while the tested vehicle is driven at various speeds and loads on a dynamometer for 240 seconds. Another transient load test is the **BAR31,** a 31 second test cycle that includes an acceleration ramp similar to the IM240.

Malfunction Indicator Lamp (MIL) A lamp on the instrument panel that lights when the PCM detects an emission-related problem, similar to a "CHECK ENGINE" light.

Manifold Absolute Pressure (MAP) The pressure in the intake manifold referenced to a perfect vacuum. Since manifold vacuum is the difference between manifold absolute pressure and atmospheric pressure, all the vacuum readings in the composite vehicle are taken at sea level, where standard atmospheric pressure equals 101 kPa or 29.92 in. Hg.

Mass Air Flow (MAF) System A fuel-injection system uses a MAF sensor to measure the mass (weight) of the air drawn into the intake manifold, measured in grams per second.

On-Board Diagnostics (OBD) A diagnostic system contained in the PCM, which monitors computer, inputs and outputs for failures. OBD-II is an industry-standard, second generation OBD system that monitors emission control systems for degradation as well as failures.

Pulse Width Modulation (PWM) An electronic signal with a variable on-off time.

Powertrain Control Module (PCM) The electronic computer that controls the engine and transmission; similar to an ECM, VCM, ECA, ECU, or SBEC.

Root Cause of Failure A component or system failure which, if not repaired, can cause other failures. If the secondary failure is repaired, but the root cause is not repaired, the secondary failure will reoccur. For example, a plugged PCV passage can cause high crankcase pressure, resulting in leaking gaskets and seals. Replacing the gaskets and seals may stop the oil leak, but if the root cause (the PCV restriction) is not diagnosed and repaired, the oil leak will eventually return.

Scan Tool A test instrument that is used to read powertrain control system information

Scan Tool Data Information from the computer that is displayed on the scan tool, including data stream, DTC's, Freeze Frame, systems monitors, and readiness monitors.

Secondary Air Injection A system that provides fresh air to the exhaust system under controlled conditions to reduce emissions; it can be either pulse or air pump type.

Sequential Multiport Fuel Injection (SFI) A fuel-injection system that uses one electronically pulsed fuel injector for each cylinder. The injectors are pulsed individually.

Speed Density System A fuel-injection system that calculates the amount of air drawn into the engine using engine rpm, air temperature, manifold vacuum, and volumertic efficiency, rather than measuring the mass or volume of air directly with an airflow meter.

Three-Way Catalytic Converter (TWC) A catalytic converter system that reduces levels of HC, CO, and NO_x.

Trip A driving cycle that allows an OBD-II diagnostic test (monitor) to run.

Notes

Notes

Notes

Notes

Notes

Notes

Notes

Notes

Notes